Microbial Diseases of
Marine Organisms

Microbial Diseases of Marine Organisms

Editor

Snježana Zrnčić

MDPI • Basel • Beijing • Wuhan • Barcelona • Belgrade • Manchester • Tokyo • Cluj • Tianjin

Editor
Snježana Zrnčić
Croatian Veterinary Institute
Zagreb
Croatia

Editorial Office
MDPI
St. Alban-Anlage 66
4052 Basel, Switzerland

This is a reprint of articles from the Special Issue published online in the open access journal *Journal of Marine Science and Engineering* (ISSN 2077-1312) (available at: https://www.mdpi.com/journal/jmse/special_issues/cynthia_microbial_diseases_marine_organisms).

For citation purposes, cite each article independently as indicated on the article page online and as indicated below:

LastName, A.A.; LastName, B.B.; LastName, C.C. Article Title. *Journal Name* **Year**, *Volume Number*, Page Range.

ISBN 978-3-0365-7236-9 (Hbk)
ISBN 978-3-0365-7237-6 (PDF)

Contents

About the Editor

Snježana Zrnčić

Snježana Zrnčić, PhD, Doctor of Veterinary Medicine, works in the National Reference Laboratory for Fish and Shellfish Diseases and has more than 30 years of experience in the pathology of aquatic organisms. Her general interests lie in the study of diseases among farmed marine fish and bivalve mollusks. She has participated in several research projects on the subject. Currently, she is serving as the General Secretary of the European Association of Fish Pathologists (EAFP). Furthermore, she participated in organization of conferences organized by the EAFP and also as a program co-chair of the European Aquaculture Society conference. She served as external expert for the UN FAO in the field of diagnostics. During her career she published more than 50 peer reviewed papers and more than 60 conferences paper, chapters in five books and edited three thematic manuals. She mentored master and doctoral thesis and participated in several working groups for advices to decision making bodies nationally and on European level.

Journal of
Marine Science and Engineering

Editorial

Microbial Diseases of Marine Organisms

Snježana Zrnčić

Laboratory for Fish Pathology, Croatian Veterinary Institute, 10000 Zagreb, Croatia; zrncic@veinst.hr

Citation: Zrnčić, S. Microbial Diseases of Marine Organisms. *J. Mar. Sci. Eng.* **2022**, *10*, 1682. https://doi.org/10.3390/jmse10111682

Received: 1 November 2022
Accepted: 2 November 2022
Published: 7 November 2022

Publisher's Note: MDPI stays neutral with regard to jurisdictional claims in published maps and institutional affiliations.

Healthy oceans and marine environments provide critical life support functions upon which human health and well-being depend [1]. Multiple benefits are derived from marine and coastal ecosystem at local, regional and global scales, ranging from pollution control, storm protection, shoreline stabilization and habitats for species to climate mitigation and food provisioning.

At present, we are facing increasing threats to the sustainability of the marine environment caused by industrialization, tourism, marine traffic and global warming. Marine organisms, whether they are prey or predators in the food chain, are very important members of the marine ecosystem. They form associations with microorganisms, including protists, bacteria, fungi and viruses, and their relationships are mostly mutually beneficial symbiont systems. However, environmental changes induced by anthropogenic impact and climate changes may alter the symbiont relationship, and microorganisms could influence the health, physiology, behaviour and ecology of marine animals [2]. Over time, many different pathogenic microorganisms have been reported as causes of mortality of fish, molluscs, crustaceans and other marine organisms. These disease outbreaks may lead to a large decline in the host population, resulting in the endangerment of the affected species and causing an imbalance in the marine environment.

This Special Issue, titled "Microbial Diseases of Marine Organisms", is conceived as a contribution to the knowledge of the deleterious impacts of microorganisms on marine fish, molluscs, crustaceans, cetaceans or other organisms. Data presented in this Special Issue provide valuable information which could be used for establishing strategies for disease mitigation and control and consequently contribute to preserving the sustainability of the marine ecosystem. A variety of hosts such as corals [3]; different species of endemic, farmed or commercial bivalve mollusc species [4–6]; decapods [7]; and wild fish in public aquaria [8] and aquaculture facilities [9] endangered by different microorganisms is presented in this issue.

We have learnt that coral reefs are among the most biodiverse biological systems on Earth [3]. They are classified as marine invertebrates and filter the surrounding food and other particles in seawater, including pathogens such as viruses. Viruses act as both pathogens and symbionts for metazoans. Marine viruses, which are abundant in the ocean, are mostly single- or double-stranded DNA and single- or double-stranded RNA viruses. These findings were obtained using advanced identification methodologies to detect the presence of viruses in coral reefs; PCR analyses, metagenomic analyses, transcriptomic analyses; and electron microscopy. The review paper in this issue discusses and presents the discovery of different viruses in the marine environment and their hosts, the viral diversity in coral and also the presence of viruses in corallivorous fish communities in reef ecosystems. The detection methods were described, as well as the occurrence of marine viral communities in marine sponges. It was concluded that marine viral communities play a crucial role in biogeochemical cycles in the ocean, indirectly and directly. These viral communities induce mortalities and diseases in the reef ecosystem through abiotic and biotic factors, which cause disturbances in the symbiotic relationship between the coral and their surrounding hosts. The review emphasizes marine viruses from the ocean, coral-associated viruses and marine sponge and coral fish viruses in reef ecosystems, examining previous research via traditional methods to modern advanced approaches. The findings of this review have enhanced our understanding of coral–virus interactions and enriched our understanding of reef-associated virus interactions and the diversity of viral communities in marine environments.

Other studied species were red claw crayfish (*Cherax quadricarinatus*) and red swamp crayfish (*Procambarus clarkia*), important for food and ornamental purposes, imported to South Korea from China and Indonesia [7]. The research comprised PCR testing to detect infectious hypodermal and hematopoietic necrosis virus (IHHNV or Decapod penstylhamaparvovirus 1). IHHNV was detected in tissue samples pooled from nine out of ten batches of red claw crayfish imported from Indonesia. Phylogenetic analysis of PCR amplicons from representative pools clustered the IHHNV strain with infectious-type II sequences commonly detected in Southeast Asian countries, rather than with type III strains detected previously in white leg shrimp (*Penaeus vannamei*) cultured in South Korea. IHHNV DNA was detected most frequently in the muscle, followed by hepatopancreas and gill tissues, suggesting that red claw crayfish could be a potential carrier of the virus. It was concluded that transboundary movements can cause significant environmental disturbance and provide opportunities for the inadvertent translocation of disease-causing pathogens to new locations. The detection of IHHNV infection in *C. quadricarinatus* by identifying the viral DNA in tissue samples of the commodity imported into South Korea as the IHHNV type II strain indicates that red claw crayfish could be a potential carrier of the infectious IHHNV, posing a threat to the cultured and wild population of crustaceans in South Korea.

The most represented marine species of this issue were bivalve molluscs: from farmed European flat oysters (*Ostrea edulis*) [5], followed by farmed and free-living Portuguese oysters (*Crassostrea angulata*) and Pacific oysters (*Crassostrea gigas*) [6], to the critically endangered noble pen shell (*Pinna nobilis*) endemic to the Mediterranean area [4]. Research on flat oysters describes the first occurrence and molecular identification and epidemiology of parasites from the genera *Bonamia*, phylogenetically positioned into a clade microcell within the genus *Haplosporidia* in a farming area along the Croatian Adriatic Coast [5]. PCR analysis and sequencing for SSU rDNA gene and BLAST analysis confirmed infection with *Bonamia exitiosa*. Although prevalence in a five-year period ranged from 3.3 to 20% at the different sites, there were no mortalities reported from the infected sites, and it seemed that infection of flat oysters with *B. exitiosa* did not affect their health. Attempt to prove the Pacific oyster as a putative vector of the parasite failed. The phylogenetic analysis did not disclose any information on the source of *B. exitiosa* origin. Since the Croatian isolate showed 100% similarity to previously sequenced isolates from Chile or Australia based on the SSU rDNA gene, the sequencing of additional genes or the whole genome should be carried out to provide us with more details on the phylogeny of the Croatian isolates. More comprehensive molecular studies of the *B. exitiosa*, together with an investigation of the natural population of *O. stentina*, which are susceptible species from production areas and natural beds along the Eastern Adriatic coast, could confirm the natural-historical origin of the parasite *B. exitiosa*. Other studied bivalve molluscs species are species nationally important for Portugal; Portuguese oyster (*Crassostrea angulata*) and Pacific oyster (*Crassostrea gigas*) from four distinctive areas in Portugal were studied to evaluate their sanitary status [6]. Collected Pacific oyster populations were cultivated in a strong ocean-influenced environment, and Portuguese oyster populations were cultivated in wild beds. The histopathological examination of both oyster species revealed the presence of parasites in gills, mantle epithelium, digestive gland tubules and connective tissue, with a moderate prevalence. In both populations, hemocytosis was observed in the connective tissue, oedema and metaplasia in the digestive gland and necrosis in the tissues. In wild populations from the Sado and Mira estuaries, the prevalence of mud blisters and gill lesions was higher than from populations produced on 0.50 m tables from mudflats. It was concluded that diseases are important risk factors which are caused, in many cases, by non-compliance with basic management rules, namely, the level of animal load in production areas, the length of time the bivalve molluscs remain in these areas, and the introduction of seeds of unknown origin. Effective biosecurity measures and correct and early diagnostic techniques are essential to control pathogenic agents. In the production areas of the Aveiro and Alvor lagoons, these measures are implemented by producers and by the authorities. Producers understand that it is essential to prevent mortalities in bivalve mollusc populations, namely, avoiding overcrowding and preventing diseases while in Sado and Mira estuaries, wild populations were proven to be more susceptible to lesions than oysters produced on tables under the supervision of producers. Differently from previous research, another study is dedicated to the noble pen shell (*Pinna nobilis*), the largest bivalve (60–120 cm), endemic to the Mediterranean Sea [4]. It is an inhabitant of shallow waters along the Croatian Adriatic coastline and suffered high mortalities similar to other parts of the Mediterranean. The results of the study presented in this Special Issue contribute a description of the diagnostics of causative agents of mortalities and epidemiology in Mljet National Park and the Northern Adriatic. It seems that mortalities were caused by infection with the haplosporidian parasite *Haplosporidium pinnae* and bacterium *Mycobacterium* sp. The spreading pattern of the mortalities based on a pilot study undertaken in Mljet National Park, an area with a dense population of noble pen shells, was evaluated. The results of the study support the hypothesis that the increase in mortalities was influenced by high temperatures, as peak mortalities in the studied area occurred in August with a sea temperature of more than 26 °C. In addition, multifactorial causality was proven, as the presence of *Mycobacterium* sp. alone was detected a long time before mortalities occurred, but coinfection of *Haplosporidium pinnae* was also detected after mortality. Still, a multidisciplinary experts' approach is needed to explain the phenomenon and to set up an efficient program for the protection of noble pen shell from extinction.

The focus of the next two papers are marine fish: one studied lesser-spotted dogfish (*Scyliorhinus canicula*) juveniles reared in public aquaria which suffered from infection with two different species from the genus *Vibrio* [8], and the other studied the skin microbiota of farmed fish correlated to the use of antibiotics [9]. Although elasmobranchs are endangered species in the Mediterranean Sea, classified as on the decline due to habitat degradation and consequent to the direct impacts of fishing, lesser-spotted dogfish, a small demersal shark, is classified as being of least concern (LC) by the IUCN. Its diet and habitat requirements, as well as easy reproduction in captivity, make them favourable species in public aquaria. Reports on infectious diseases affecting sharks are scarce, and therefore, research on their susceptibility to different infectious agents may contribute to the mitigation of elasmobranchs' decline. This research proved susceptibility to two *Vibrio* species: *V. crassostreae*, previously described as a pathogen of molluscs and fish, and *V. cyclotrophicus*, reported in molluscs and as a member of the microbiome of marine copepods [8]. This study described suitable diagnostic methodology to identify specifically *V. crassostreae* and *V. cyclitrophicus*, underlining the criticalities in the identification process concerning the techniques adopted (API® 20E, MALDI-TOF MS, molecular biology) and the need for developing specific guidelines for the identification of non-major pathogenic *Vibrio* species. In addition, it highlighted the need for in-depth studies of the pathogenic mechanism of these bacteria in sharks. Bacteria from the genus *Vibrio*, together with those from the genus *Pseudomonas*, were prevalent in the study of skin microbiota of farmed European seabass (*Dicentrarchus labrax*) [9]. Some of the microbiota that were identified are known to be pathogenic to fish: *V.alginolyticus*, *V. anguillarum* and *V. harveyi*. *Vibrio* strains showed higher resistance to studied antibiotics compared to previous studies. This study provides, for the first time, information on the cultivable skin bacteria that were associated with healthy European seabass under culture conditions with and without the use of antibiotics. The obtained information will be useful in assessing how changes in cultivable microbiota may affect the health of farmed European seabass, indicating a potential problem for fish health management during disease outbreaks. Interestingly, some resistant bacteria were detected among isolated bacteria, a component of skin microbiota from the farm without the use of antibiotics that raised many different questions on the influence of the fish farms on the environment but also the influence of the environment and anthropogenic activity on fish farming. The presence of resistant microorganisms could potentially endanger consumers' health and contribute to horizontal gene transmission. However, there is little data on the antimicrobial resistance gene transmission (AGR) in the marine environment and future studies should put a lot of effort into elucidating pathways and possibilities of AGR spreading.

Each of the published articles in this Special Issue tackled one of the niches within the marine environment and fulfilled a strong need for further and more detailed research on different marine organisms and their interactions with different microbes and the outcomes of these interactions.

Funding: This research received no external funding.

Conflicts of Interest: The author declares no conflict of interest.

References

1. OECD. OECD Work in Support of Sustainable Ocean. Brochure. 2022. Available online: https://www.oecd.org/environment/2022-OECD-work-in-support-of-a-sustainable-ocean.pdf (accessed on 18 July 2022).
2. Aprill, A. Marine Animal Microbiomes: Toward Understanding Host–Microbiome Interactions in a Changing Ocean. *Front. Mar. Sci.* **2017**, *4*, 222. [CrossRef]
3. Ambalavanan, L.; Iehata, S.; Fletcher, R.; Stevens, E.H.; Zainathan, S.C. A Review of Marine Viruses in Coral Ecosystem. *J. Mar. Sci. Eng.* **2021**, *9*, 711. [CrossRef]
4. Mihaljević, Ž.; Pavlinec, Ž.; Zupičić, I.G.; Oraić, D.; Popijač, A.; Pećar, O.; Sršen, I.; Benić, M.; Habrun, B.; Zrnčić, S. Noble Pen Shell (*Pinna nobilis*) Mortalities along the Eastern Adriatic Coast with a Study of the Spreading Velocity. *J. Mar. Sci. Eng.* **2021**, *9*, 764. [CrossRef]
5. Oraić, D.; Beck, R.; Pavlinec, Ž.; Zupičić, I.G.; Maltar, L.; Miškić, T.; Acinger-Rogić, Ž.; Zrnčić, S. *Bonamia exitiosa* in European Flat Oyster (*Ostrea edulis*) on the Croatian Adriatic Coast from 2016 to 2020. *J. Mar. Sci. Eng.* **2021**, *9*, 929. [CrossRef]
6. Pires, D.; Grade, A.; Ruano, F.; Afonso, F. Histopathologic Lesions in Bivalve Molluscs Found in Portugal: Etiology and Risk Factors. *J. Mar. Sci. Eng.* **2022**, *10*, 133. [CrossRef]
7. Lee, C.; Choi, S.-K.; Jeon, H.J.; Lee, S.H.; Kim, Y.K.; Park, S.; Park, J.-K.; Han, S.-H.; Bae, S.; Kim, J.H.; et al. Detection of Infectious Hypodermal and Hematopoietic Necrosis Virus (IHHNV, Decapod Penstylhamaparvovirus 1) in Commodity Red Claw Crayfish (*Cherax quadricarinatus*) Imported into South Korea. *J. Mar. Sci. Eng.* **2021**, *9*, 856. [CrossRef]
8. Tomasoni, M.; Esposito, G.; Mugetti, D.; Pastorino, P.; Stoppani, N.; Menconi, V.; Gagliardi, F.; Corrias, I.; Pira, A.; Acutis, P.L.; et al. The Isolation of *Vibrio crassostreae* and *V. cyclitrophicus* in Lesser-Spotted Dogfish (*Scyliorhinus canicula*) Juveniles Reared in a Public Aquarium. *J. Mar. Sci. Eng.* **2022**, *10*, 114. [CrossRef]
9. Ramljak, A.; Vardić Smrzlić, I.; Kapetanović, D.; Barac, F.; Kolda, A.; Perić, L.; Balenović, I.; Gavrilović, A. Skin Culturable Microbiota in Farmed European Seabass (*Dicentrarchus labrax*) in Two Aquacultures with and without Antibiotic Use. *J. Mar. Sci. Eng.* **2022**, *10*, 303. [CrossRef]

Journal of
Marine Science and Engineering

Review

A Review of Marine Viruses in Coral Ecosystem

Logajothiswaran Ambalavanan [1], Shumpei Iehata [1], Rosanne Fletcher [1], Emylia H. Stevens [1] and Sandra C. Zainathan [1,2,*]

[1]   Faculty of Fisheries and Food Sciences, University Malaysia Terengganu,
     Kuala Nerus 21030, Terengganu, Malaysia; loga_5427@yahoo.com (L.A.);
     shumpei@umt.edu.my (S.I.); rosannefletcher90@gmail.com (R.F.); emylia.stevens@gmail.com (E.H.S.)
[2]   Institute of Marine Biotechnology, University Malaysia Terengganu,
     Kuala Nerus 21030, Terengganu, Malaysia
*   Correspondence: sandra@umt.edu.my; Tel.: +60-179261392

Abstract: Coral reefs are among the most biodiverse biological systems on earth. Corals are classified as marine invertebrates and filter the surrounding food and other particles in seawater, including pathogens such as viruses. Viruses act as both pathogen and symbiont for metazoans. Marine viruses that are abundant in the ocean are mostly single-, double stranded DNA and single-, double stranded RNA viruses. These discoveries were made via advanced identification methods which have detected their presence in coral reef ecosystems including PCR analyses, metagenomic analyses, transcriptomic analyses and electron microscopy. This review discusses the discovery of viruses in the marine environment and their hosts, viral diversity in corals, presence of virus in corallivorous fish communities in reef ecosystems, detection methods, and occurrence of marine viral communities in marine sponges.

Keywords: coral ecosystem; viral communities; corals; corallivorous fish; marine sponges; detection method

Citation: Ambalavanan, L.; Iehata, S.; Fletcher, R.; Stevens, E.H.; Zainathan, S.C. A Review of Marine Viruses in Coral Ecosystem. *J. Mar. Sci. Eng.* **2021**, *9*, 711. https://doi.org/10.3390/jmse9070711

Academic Editor: Snježana Zrnčić

Received: 7 May 2021
Accepted: 25 June 2021
Published: 27 June 2021

Publisher's Note: MDPI stays neutral with regard to jurisdictional claims in published maps and institutional affiliations.

1. Coral Ecosystem

The earth is covered by an ocean that contains about 97% of the planet's water [1]. Mora et al. [2] have estimated that about 2.2 ± 0.18 million species are found in the marine environment but only 91% of the ocean species have yet been discovered. The coral ecosystem consists of sponges, coral, corallivorous, crustaceans, molluscs and other organisms that live beneath the reef ecosystem. Sponges are among the oldest metazoans which can be divided into four classes: Hexactinellida, Calcarea, Demospongiae and Homoscleromorpha [3]. Sponges feed on plankton that include bacteria, algae, protozoa and microscopic animals which transfer carbon flow to higher trophic levels [4]. These sponges are also known to be food sources for organisms, for example, fish, crustaceans, sea urchins, star fish and molluscs [4]. Sponge's morphology is very diverse with a colorful array ranging from amorphous types, to branching, with a great variety in length and size [5]. Moreover, sponges are plentiful and functionally essential for coral reef systems [6]. They play a crucial role in numerous ecosystems such as substrate accretion [7] and erosion [8,9]. As described by Aerts [10], out of 128 sponge species, 30 interact with corals in coral over-growth [9]. As reviewed by Bell [6], functional roles that are played by sponges in Caribbean coral include increasing coral survival by binding live corals to reef frame and preventing entry to their skeletons by excavating organisms, nutrient cycling, bioerosion reworking of solid carbonate, primary production via microbial symbionts, removing prokaryotic plankton at water column and providing food sources for organisms.

Corallivores are known as fishes that feed on live corals in reef ecosystems. These coral-feeding fishes have shown that almost 80% of their feeding is based on coral, which assumes that they are dependent on coral for their survival [11]. They can be classified as polyps-feeders, removing coral tissues. Persistent predation caused by coral fish aggravate the effects of coral disturbance and slow down reef recovery [11]. As reviewed by Cole [11], the corallivorous species can vary among habitats and geographic areas. Coral colonies as a food source for these corallivores need to find a balance between feeding severity and coral regeneration [11].

Meanwhile, coral reefs are among the most biodiverse biological systems on the planet, the persistence of which relies on the reef-building capacity of scleractinian corals. The reef-building corals accrue almost 250 million years with hundreds of scleractinian coral species that exhibit in multiple colony sizes, shapes and life spans, providing a wide assemblage of territory for molluscs, fish and crustaceans [12]. Corals are made up of networks of polyps and gastrovascular system that consist of tissue layers that contain numerous cell types such as epidermis mesoglea, gastrodermis, coelentron, gastrodermis mesoglea and calicoblastic epithelium [12]. According to Tresguerres and Barott [13], the corals' oral ectoderm connects to the external environment and is involved in the production of the mucus layer which helps corals in capturing prey and guarding against nematocysts. Hence, corals filter the surrounding food particles and other particles in seawater including pathogens such as viruses.

Corals are either hermatypic (reef-building) or ahermatypic (non-reef-building). Hermatypic corals flourish in shallow waters which contain millions of zooxanthellae. *Symbiodinium* are colloquially known as zooxanthellae. Due to the complex beneficial interactions among coral hosts, unicellular algae and dinoflagellates *Symbiodinium* spp., and their different microbiomes, coral reefs flourish in oligotrophic tropical waters [14]. Coral beneficial interaction with *Symbiodinium* algae permits the coral to harness power from daylight (photosynthesis), as fixed natural carbon is moved to the host while the algae obtain inorganic supplements reprocessed from the host's metabolism, for example ammonium and carbon dioxide [15]. These resources supply the coral with vitality for growth, reproduction and respiration [14]. Zooxanthellae conducts photosynthesis from sunlight as it captures and transfers 95% of the energy generated to the coral polyps. This relationship between zooxanthellae and corals supports both the reef and the zooxanthellae, sustained itself by nutrients such as inorganic substances and acquiring refuge in exchange. Global climate change can also stress the coral, collapsing the symbioses, which leads to bleaching (paling) or loss of zooxanthellae [14].

Cziesielski et al. [16] discovered the first mass coral bleaching in 1988 and over the past 4 years alone prominent coral reef ecosystems such as the Great Barrier Reef have lost about 50% of shallow-water corals [17]. Coral reefs depend on a differing consortium of free-living and host-related microorganisms for the capture, maintenance, and reuse of nutrients and minor components that permit these environments to flourish in the marine environment, which can be compared to a desert [18,19]. Climate change can trigger disruption of the natural microbiome which could lead to a state of dysbiosis with the exposure of opportunistic and probably pathogenic taxa, resulting in an inclining incidence of disease, bleaching, and host mortality [20,21]. It has been proposed that the microbiome could play a central role in coral reef reclamation, including marine viruses [21]. These changes in the microbial community cause the aggravations that could act as early warning signals, since these movements may anticipate visual indications of bleaching and tissue necrosis [22–25]. Coral reefs have endured uncommon losses and are critically threatened by continuous elevation in ocean surface water temperature and ocean acidification due to climate change, despite viral infection in corals and their worldwide significance [26,27].

According to Thurber et al. [28], the coral holo-biont contains diverse viral-like particles (VLPs) which were first observed in stony corals (cnidarians) and the sea anemone [29]. It is reported that larger viruses could infect dinoflagellate endosymbionts of the corals due to environmental factors. Virus-associated corals infect the coral tissues and the coral surface microlayers (CSM) [30–33]. Cram et al. [34] predicted that high viral infection is the source of marine bacterial mortality that affects shape together with structure of the microbial host community in the marine ecosystem and transfers genetic material between microorganisms, known as horizontal gene transfer (HGT) [35]. Viruses also encode auxiliary metabolic genes (*AMGs*) that increase microbial host growth and fitness [36–38]. Viruses exist as both pathogen and symbiont of metazoans, controlled by the environmental factors and mutual relationship between the host and virus itself [39,40]. The ecological roles of the virus communities in health and disease in corals are still inadequately recognized despite the availability of advanced molecular technology. According to Thurber et al. [28], the stress-effect of viral infections damages the coral tissues that drive carbon from the benthos into the water column in the reef ecosystem. Marine bacterioplankton blooms trigger viral production in reef waters that releases more carbon and nitrogen into the environment. These occurrences induce viral production and mortality in corals and lead to coral reef decline [28]. Generally, corals are involved in the biogeochemical cycles in the reef ecosystems due to the release of dissolved organic carbon (DOC) [28]. The presence of DOC stimulates bacterial growth and it has been reported to enhance the DOC flux processes, and nutrients released from CSM have negative feedback on the coral reefs [41,42]. A shift in the relative abundances of eukaryotic viruses and phages has been reported by Correa et al. [43] (lab-stressed corals, Hawaii and Florida) and Soffer et al. [44] (bleached, disease and healthy corals, Caribbean reefs). The shifts in the relative abundance of virus families may also play a role in coral mortality and disease. For instance, the Circoviridae family is one of the most abundant viruses in corals that causes a disease known as the white plague [28].

According to Sweet et al. [45], viral agents may act in various manners, including critical infection, reactivation of inert disease, immune suppression (declining of the immune system of the host or weakening of the immune system due to natural aging), where it is able to destroy algae symbionts under environmental pressure and may participate in coral bleaching reactions. Nonetheless, some of these viruses are possibly host-specific to the symbiotic algae and may perform as primary pathogens in coral disease. In conjunction with these known groups, various viral sequences have been distinguished that could nt be appointed to any known families of viruses, demonstrating that the coral microbiome is a rich environment for novel viral revelations [45].

The bacteriophage-adhering-to-mucus (BAM) model had been hypothesized and named due to the observation of enrichment of phage in mucus occurring through interaction of mucin glycoproteins and Ig-like proteins that realm on phage capsids [45]. These BAM models assimilate a mechanism in supporting the specific relationship of mutualistic bacteria with host, and dispenses an evolutionary structure for a process of specific co-advancement of a phage-bacterial-host related microbiome. Therefore, evidence has demonstrated the role of viruses in the control of coral-related bacterial communities, where viral lysis rates rise in the surface mucus layer and complex associations are presented in the bacteriophage-adhering-to-mucus model. Viruses were also indirectly involved in controlling pathogenic bacterial populations and disease pervasiveness [45]. Wood-Charlson et al. [46] showed that diverse viral families had been detected from coral and symbiotic microbes by metagenomics. They reported that these viruses are likely to play numerous, commensal roles and are parasitic, regarding the health of coral reefs. Regardless of these assorted varieties, a number of taxonomic classifications are generally found in corals, including bacteriophages belonging to an order of the Caudovirales, and eukaryotic nucleocytoplasmic large DNA viruses (NCLDVs) associated with families such as Phycodnaviridae, Mimiviridae, Poxviridae and Iridoviridae, as well as Polydnaviridae and Retroviridae [43,46]. Other than the families that affect unicellular algae, there are few a number of coral-associated families known to infect plants and/or fungi and protists including Geminivirdae, Nanoviridae, Tymoviridae, Potyviridae, Tombusviridae, Caulimoviridae, Alphaflexiviridae, Endornaviridae, Partitiviridae and Reoviridae [46].

2. Progress of Marine Virus Research

Marine viruses are found to be abundant in the ocean in many organisms as they play vital roles in global geochemical cycles [47]. Viruses are widely distributed in various environments including extreme environments, such as hydrothermal vents, cold springs and hypoxic saline conditions. They exist widely in seawater [48] and sediment [49].

Spencer et al. [50] discovered the first bacteriophage from the marine environment. Viruses recycle nutrients together with organic matter via a process known as the viral shunt. The cellular materials released as particular or dissolved organic material are not directly available for utilization by organisms from higher trophic levels but are primarily utilized by predominantly heterotrophic bacteria, although some efforts have shown nutrients released in this manner are rapidly assimilated by eukaryotic plankton [51]. Spatio-temporal dynamics of these viruses accounted for 10% of phytoplankton mortality within the course of phytoplankton blooms and additionally stimulated recycling of nutrients, together with organic matter, via viral shunt [52–54]. Research advances in these environments have encountered marine viruses and their hosts and organisms including plankton and aquatic invertebrates.

The Kill the Winner (KtW) hypothesis proposed that viruses are sustained via host specific infection including plankton and lysis, the most abundant microorganisms in the environment [55]. Numerous efforts have been made to identify and isolate viruses from cultivated microorganisms and have increased understanding of genomic structure and the host range of marine viruses, especially from Cyanobacteria (such as *Prochlorococcus* and *Synechococcus*) [56–59]. The RNA-dependent RNA-polymerase (*RdRp*) gene is an essential protein encoded in the genomes of all RNA-containing viruses such as Picorna-like viruses and DNA containing viruses such as T4-like myoviruses (gene marker g23 and g20) have limited the research on viral diversity to genome fingerprinting [60–63]. Thereby, metagenomics studies were introduced, to overcome the bottleneck of cultivation and the lack of universal markers, in the early twenty-first century. Breitbart et al. [64] reported that both single-gene-based and genome-fingerprinting-based methods have revealed the temporal and spatial dynamics of marine viral communities.

Metagenomics-based studies have enabled scaling of the information on viral genomics and sequencing of the fragmented nucleic acid from seawater and marine sediments [65]. However, several important issues, despite the progress in the study of of marine viruses, remain unexplored, for instance, the examination of specific-host interactions, expansion of spatio-temporal marine viral studies, linking the environmental viruses with their hosts and enhancement of knowledge particularly regarding deep-sea viral communities and viral auxiliary metabolic genes (AMGs) [38]. Advanced informatics and theoretical research have unveiled the biological basis of complex host range patterns and explicated largely unknown viral sequences in marine ecosystems [66,67]. Regarding this, the progress of marine viruses from marine sponges are recommended so that we know the abundance or presence of viral communities that have been detected, making it easier for other researchers to refer to.

3. Marine Viruses and Their Host

The major impact of viruses in the marine environment began when it was discovered that seawater contains around 10^{10} viruses per liter [68]. In the mid-1970s, viruses or virus-like particles (VLPs) were reported in numerous taxa of algae. In the 1980s, a group of large double stranded(ds) DNA-containing viruses (Chloroviruses) was discovered. These viruses infect and replicate in unicellular, eukaryotic, symbiotic, chlorella-like green algae known as Micratinium (formerly known as Zoochlorellae) (Chlorophyta). As described by Breitbart et al. [69], viral abundances are known to be tightly linked to their host and are generally more abundant in phytoplankton and Cyanobacteria than in any other microorganisms.

In the past, large-scale spatial investigations of viral dispersion in the Pacific and South Atlantic Oceans have utilized flow cytometry to reveal that viral abundance in 200 m of the water surface column is high in tropical and subtropical districts yet lower in Antarctic waters [70,71]. A virus "hot-spot" was recognized in the mid-scope region of the North Pacific supporting the hypothesis that large-scale distribution patterns of viruses are influenced by host distributions and physical procedures [71]. Brum et al. [72] used cultivation-independent methods such as metagenomics and revealed that the Southern Ocean is dominated by lysogenic viruses, which are termed as "seasonal time bombs" since they can shift to a lytic cycle as their bacterial host production rises. Roux et al. [73] collected virome data that identified virally encoded AMGs (Auxiliary Metabolic Genes) in surface waters that appear to be involved in nitrogen and sulfur cycling.

The Pacific Ocean Virome (POV) [74] is another significantly important curated dsDNA virome dataset that comprises 32 quantitatively representative viromes collected from different depths and seasons in transects from coastal waters to open-ocean waters in the Pacific Ocean [74]. POV study revealed that the richness of viruses in the Pacific reduced from deeper to surface waters, from winter to summer, and, in surface layers, with distance from the shore. Other than that, data collected by the TARA ocean expedition were used to develop a global virome map of dsDNA viruses sampled in surface and deep-ocean waters [73], along with the genomic data from the Malaspina 2010–11 Circumnavigation Expedition, which has assessed pelagic processes along the Indian, Pacific, and Atlantic Oceans [75]. The authors found that 38 of 867 viral clusters were locally or globally abundant and represented almost 50% of the viral population in any given Global Ocean Virome (GOV). Such huge datasets are critical in developing our insights into marine viruses on a global scale and are important in divulging uncultivated novel viruses and initiating important marine ecosystem models.

Hence, the International Committee on Taxonomy Viruses (ICTV) have classified and named viruses according to their taxonomy and binomial species [76]. The key factors that are used for the identification and classification of viruses include the display of capsid protein and viral morphology via TEM [77]. Nineteen families of unassigned viruses were reported, whereas approximately 5000 marine viruses were assigned to 26 families according to ICTV [67].

Moreover, viruses can be grouped based on their hosts such as bacteria, animal, archaea, algae and plant viruses (Table 1). Authors have also discovered single stranded DNA (ssDNA) and double stranded RNA (dsRNA) viruses other than the double stranded DNA (dsDNA), abundant in the ocean with marine organisms as their hosts.

Table 1. Marine viruses and their hosts in the marine environment adapted from He et al. [67] and King et al. [78].

Virus Genome	Order	Family	Morphology	Size (nmin Diameter)	Host Species	References
Double stranded DNA (dsDNA)	Caudovirales	Myoviridae	Polygonal head (icosahedral) with contractile tail (helical)	50–110	Bacteria	
	Caudovirales	Podoviridae	Icosahedral	130–200	Bacteria	[79–81]
	Caudovirales	Siphoviridae	Icosahedral with noncontractile tail	60	Bacteria	
	Herpesvirales	Herpesviridae	Pleomorphic, icosahedral, enveloped	150–200	Fish, corals, mammals, mollusks, and bivalve	[82]
	Ligamenvirales	Lipothrixviridae	Thick rod with lipid coat	400	Archaea	[83]
	Unassigned	Baculoviridae	Enveloped rods, some with tails	65–100 × 230–335	Crustaceans	[84]
	Unassigned	Corticoviridae	Icosahedral with spike	60–75	Bacteria	[85]
	Unassigned	Iridoviridae	Round, icosahedral	190–200	Fish, Mollusks	[86]
	Unassigned	Mimiviridae	Icosahedral with microtubule-like projections	650	Marine fish	[87]
	Unassigned	Nimaviridae	Enveloped, ovoid with tail-like appendage	275	Marine crustaceans	[88]
	Unassigned	Papillomaviridae	Round, icosahedral	40–50		[89]
	Unassigned	Phycodnaviridae	Pleomorphic, icosahedral, enveloped	130–200	Algae	[90]
	Unassigned	Tectiviridae	Icosahedral, with noncontractile tail	60	Bacteria	[91,92]
	Unassigned	Totiviridae	Round, icosahedral	30–45	Protist	[93]
		Anelloviridae	icosahedral, circular	30	Vertebrates	
		Circoviridae	icosahedral, circular	12–27	Vertebrates	
		Geminiviridae	icosahedral	22 × 38	Protozoa	
Single stranded DNA (ssDNA)	Unassigned	Inoviridae	Genus *Inoviruses*: filamentous; plectroviruses: rod-shaped	inoviruses: 7 × 700–3500; plectroviruses: 15 × 200–400	Bacteria	[78]
	Unassigned	Microviridae	icosahedral	25–27	Bacteria	
	Unassigned	Nanoviridae	icosahedral	18–20	Protozoa	
	Unassigned	Parvoviridae	icosahedral, linear segment	21–26	Vertebrates, invertebrates	
	Bunyavirales	Peribunyaviridae	Round, enveloped	80–120	Crustaceans	[94]
	Mononegavirales	Paramyxoviridae	Various, mainly enveloped,	60–300 × 1000	Mammals	[95]
		Rhabdoviridae	Bullet-shaped with projections	45–100 × 100–430	Fish	[96]
Single stranded RNA (ssRNA)	Nidovirales	Coronaviridae	Rod-shaped with projections	200 × 42	Crustaceans, fish, seabirds	[97,98]
		Dicistroviridae	Round, icosahedral	30	Crustaceans	
	Picornavirales	Marnaviridae	Round, icosahedral	25	Algae	[99–101]
	Picornavirales	Picornaviridae	Round, icosahedral	27–30	Algae, crustaceans, thraustochytrids, mammals	
	Unassigned	Caliciviridae	Round, icosahedral	35–40	Fish, mammals	[102]
	Unassigned	Leviviridae	Round, icosahedral	26	Bacteria	[103]
	Unassigned	Microviridae	Icosahedral with spikes	25–27	Bacteria	[104]
	Unassigned	Nodaviridae	Round, icosahedral	30	Fish	[105]
	Unassigned	Orthomyxoviridae	Round, with spikes	80–120	Fish, mammals, seabirds	[106]
	Unassigned	Togaviridae	Round, with outer fringe	66	Fish	[107]
	Unassigned	Reoviridae	Icosahedral, thick outer layers, smaller electron-dense inner cores	90–95	Algae	[108–110]
Double stranded RNA (dsRNA)	Unassigned	Birnaviridae	Round, icosahedral	60	Mollusks, fish	[111]
	Unassigned	Cystoviridae	Icosahedral with lipid coat	60–75	Bacteria	[112]
	Unassigned	Totiviridae	Round, icosahedral	30–45	Protist, shrimp	[113,114]

The concentration of free virions decreases with depth in seawater [47] with a total abundance of more than $10^{3\circ}$ [47], whereas the concentration of marine virio-plankton in surface seawater is typically billions per milliliter [54,114]. The published data on viral abundances in seawater are mostly found in freshwater, coastal water, open ocean, deep ocean and estuarine waters, but the discovery of marine viruses in Asia is still at an early stage.

Zhong et al. [115] have discovered four targeted viruses (e.g., *Cyanomyovirus*, T4-like *myovirus*, Cyanophage, Phycodnavirus) in Lake Annecy and Lake Bourget. Molecular detection such as PCR-denaturing gel gradient electrophoresis (DGGE) had been used to discover viruses in the lake water. DGGE banding pattern analysis revealed the different similarities of the viruses in both lakes. The results demonstrated the presence of 70% of *mcp* phycodnaviruses (Lake Bourget), 45% of *polB* phycodnaviruses (Lake Bourget) and 60% (Lake Annecy), respectively, 45% of psb A cyanophages (Lake Bourget and Lake Annecy), 45% of cyanomyo-viruses in Lake Bourget and 75% of T4-like myo-viruses in Lake Bourget. The total bands in common that showed the presence of the viruses in both lakes were as follows: *mcp* (85%), *polB* (54%), *g20* cyano-myovirus (52%), *psb A* cyanophages (39%) and *g23* of T4-like myo-viruses (75%), respectively. Previously, this method had focused on marine ecosystems where only a single gene marker was used typically to identify the presence of the viruses in the marine environment, *g20* [116–118], *polB* [118,119] or *mcp* [120–122]. Parvathi et al. [123] demonstrated the dynamics of virus infection of autotrophic plankton in Lake Geneva over a 5-month period and the abundances were determined via flow cytometry and PCR-DGGE method identification. The viral signature genes used by Parvathi et al. [123] were similar to those used by Zhong et al. [115] to identify the viral abundances in surface water. Parvathi et al. [123] have discovered that the abundances of picocyanobacterial hosts was in concurrence with cyanophages, which were higher in late summer.

Luo et al. [124] discovered metagenome sequence data that yielded about 16787 virus populations, and 1352 of these were identified as putative temperate phages with temperate phage marker genes. Moreover, about 12 complete archaeal virus genomes and 25 genome fragments of eukaryotic viruses have been identified [124]. They discovered that the depth of ocean inferred the Virus:Cell ratio (VC) temporal variability where, as the depth of ocean decreased, there were more temperate phages present compared to other viruses. The temperate phage peaked at a photic zone in the ocean at about 150–250 m. This peak indicates increased temperate phage productivity relative to deeper waters, with slowly growing hosts. A high VC variability driven by temporal resource variability indicates phages with more episodic virus particle production [124]. Thus, this shows that the ecology and biogeochemistry of microbial communities were impacted by viruses across the ocean.

4. Coral Viruses

Reef-building corals lay the foundation for the structure and biodiversity of the coral reef ecosystem. These huge biological structures can be seen from space and are the culmination of complex developments. The interaction between coral micro-polyps and their unicellular symbiotic algae is closely related to microorganisms such as bacteria, archaea, fungi and viruses. Reef building corals have existed in various forms for more than 200 million years, and human-induced conditions have been reported to threaten their function and durability [12]. The coral reef ecosystem is at the forefront of people's attention in the Anthropocene; for instances corals, the main species, are sensitive to human interference ranging from local activities (such as overfishing, coastal development and pollution) to worldwide phenomena (e.g., climate change and ocean acidification) [12].

Bacteria, Achaea and eukaryotic viruses are cosmopolitan and exist all over the world's oceans [125]. According to Thurber and Correa, [126], a few published studies have directly evaluated VLPs associated with shallow-water scleractinian (stony) corals, the primary architects of the coral reef ecosystem. According to Paul et al. [127], VLPs were detected in surface water, sediments and several invertebrate groups in tropical coral reefs and seagrass beds, and changes in terms of distance from coast, seasonality and salinity were recorded in the abundance of these particles. The first detection of a marine cnidarian-based VLP is based on the temperate plumose sea anemone, an old specimen of *Metridium senile* [127]. Since VLP is visualized in the tissues of the anemone, *Metridium senile* does not form symbioses with algal partners, and VLP may infect the anemone itself and/or organisms outside the anemone. Wilson et al. [128] studied the first characterisation of viruses associated with corals where VLPs were detected in both healthy and heat-stressed colonies, coral *Pavona danai, Acropora formosa* and *Stylophora pistillata*. Tail-less VLPs, hexagonal and about 40 to 50 nm in diameter, were present around the corals. A large number of VLPs were related to heat shock treatment, prompting a viral outbreak in these coral fragments. One of the most fascinating papers on coral-associated viruses in the study of VLPs was related to the coral surface microlayer of *Acropora muricata, Porites lobata* and *Porites australiensis* [129]. VLP were divided into five groups (tail phage, polyhedron/spherical, lemon-shaped, filamentous and unique VLP) and 17 subgroups based on the morphological similarity to the previous described viruses which have potential hosts existing in the coral surface microlayer, including algae, cyanobacteria, archaea, fungi and the coral animal.

A previous meta-analysis [46] found approximately 60 virus families in corals around the world as recognized by ICTV. A number of dsDNA ssDNA type viruses were identified from coral-holobiont sequence data sets and the coral transcriptomes generated from extracted holobiont RNA were dominated by 36–78% of RNA viruses that includes ssRNA viruses (1–13%) and dsRNA virus (2–31%). The analysis made by Wood-Charlson et al. [46] demonstrated that dsDNA viruses from the order Caudovirales were the only group of viruses that were dominant in all of the data sets (ICTV, 2012). Besides, most of the sequences from the coral-related viral metagenomes were dsDNA and ssDNA bacteriophages that show significant matches to the NCBI'S RefSeq Virus database.

Weynberg et al. [130] have identified that about 84% of the sequences belonged to dsDNA viruses and approximately 85.6% were ssRNA viruses from *Alternaria tenuis* (Fugi, Ascomycota). Most of the ssRNA viruses matched with the major capsid protein (MCP) gene from dinoflagellate-infecting ssRNA virus (HcRNAV) which supported the observation by Correa et al. [131] in *Montastrea cavernosa* (Animalia, Cnidaria). Studies which used approaches such as transmission electron microscopy and flow cytometry revealed the presence of VLPs associated with cultures of *Symbiodinium* (Miozoa, Dinophyceae) cells from corals. According to Thurber et al. [28], core coral viromes comprise 9–12 families in three viral lineages known as dsDNA group I, ssDNA group II and retrovirus group IV [43]. A typical virus related to the Herpesvirales order reported to be a member of the core coral virome showed 98% sequence similarities to the Herpesviridae virus [43]. Similar authors stated that only 10% of ssDNA (Circoviridae) and ssRNA (Caulimoviridae) were represented in the data set, probably due to the methodological differences among the studies. TEM-based studies identified VLP morphotypes that were comprised of enveloped, icosahedral capsids ranging between 120 to 150 nm in diameter from Kāne'ohe Bay in Hawaii in the United States, as well as in the corals that reside within a cellular vacuole alongside VLPs, and indicated its atypical herpes-like viral particles.

The morphological identification of VLPs was performed in the tissues of *Acropora muricata* colonies infected with healthy and white syndrome (WS) via TEM. This was the first study of cnidarians, which included temporal and spatial components. The characteristics of these dominant VLPs and their existence at multiple sampling time points led Patten et al. [30] to assume that the colony of the *Acropora muricata* is suffering from persistent infection of Phycodnaviridae and/or Iridoviridae. Buerger et al. [132] described the degradation of *Symbiodinium* cells and linked this to the abundance of VLPs in the coral. Table 2 shows the coral-related viruses reported in the reef ecosystem. As suggested in the case of virus-induced coral bleaching and yellow blotch disease, the virus itself causes the disease and therefore directly interacts with the coral itself, whereas the indirect processes involve bacteriophages that interact with the prokaryotic community. Phages may go through the horizontal transfer of virulence genes, increasing the virulence of the infected bacterium, which then causes coral diseases. In addition, bacteriophages may infect and lyse pathogenic bacteria, reduce the impact of disease and become a part of the coral microbiome, or reduce the external influences from the overall organism of the coral, for instance, manual application in phage therapy [132].

Table 2. Coral-associated viruses in reef ecosystems, detection method and their host.

Virus Family	Size (nm in Diameter)	Host	Coral Species	Method of Detection	References
Closteroviridae	12	Marine algae	Unclassified	Analytical fluocytometry, Tranmission Electron Microscopy	[133]
Flexiviridae, Potyviridae	200 nm–2 um	Marine algae	Unclassified	Unclassified	[134]
HaRNAV	25	Marine Algae	*Montastraea cavernosa*	Pyrosequencing	[131]
Mimiviridae, Iridoviridae	400, 120–350	Marine algae	*Porites asteroides*	Metagenome	[135]
HcRNAV	30	Marine algae	Unclassified	Transmission Electron Microscopy	[136]
Geminiviridae, Nanoviridae, Tymoviridae, Potyviridae, Tombusviridae	18–20, 30, 30, 11–20, 28	Protist, plants, invertebrate	Unclassified	Metagenome	[46]
Herpesviridae	120–150	Terestial and aquatic animals	*Montastraea annularis*, *Symbiodinium* sp., *Diploria strigosa*	Transmission Electron Microscopy, Metagenome	[43,44,137,138]
Poxviridae	200	Insects, terrestrial invertebrates (humans and birds), whales, sea lions, dolphins	*Acropora tenuis*, *Fungia fungites*, *Goniastria aspera*, *Galaxea fascicularis*, *Pocillopora acuta*, *Pocillopora damiicomis*, *Pocillopora verrucosa*	Transmission Electron Microscopy	[28,139,140]
Mimiviridae, Retroviridae, Siphoviridae, Picobirnaviridae	400, 100, 60	Eukaryotes and bacteria	*Galaxea fascicularis*, *Mycedium elephantotus*, and *Pachyseris speciosa*	Metatranscriptome and metagenome	[141]
Retroviridae, Hepadnaviridae, Parvoviridae, Iridoviridae, Herpesviridae	70–150	Target vertebrates and invertebrates	*Acanthastrea echinata*, *Diploastrea heliopora*, *Fungia* sp., and *Plerogyra sinuosa*	Metatrnascriptome and metagenome	
Herpesviridae	120–150	Bivalves, protist, bacteria	*Porites compressa*	Metagenome	[142]
Podoviridae-like, Geminiviridae-like	100	Coral surface microlayer	*Porites lobata* *Porites lutea* *Porites australiensis*	Transmission Electron Microscopy	[143]
Mimiviridae	40	*Symbiodinium* spp. cells	*Mussismilia braziliensis*	Transmission Electron Microscopy	[144]
Potyviridae	11–20 nm	*Symbiodinium* spp. cells	Unclassified	Transcriptome	[145]

Table 2. *Cont.*

Virus Family	Size (nm in Diameter)	Host	Coral Species	Method of Detection	References
Unclassified	-	*Symbiodinium* spp.	*Acropora tenuis*	Metagenomic analysis	[130]
Herpesvirus-like	120–150		*Acropora aspera, Acropora millepora*	Transmission Electron Microscopy	[41]
Viral-like particles	-	-	*Acropora muricara, Porites spp.*	Transmission Electron Microscopy	[129]
Myoviridae, Poxviridae, Microviridae	47–65, 200	-	*Acropora millepora*	Metagenome	[20]
Virus-like particles	-	-	*Acopora muricata*	Transmission Electron Microscopy	[30]
Herpesviridae, Circoviridae, Nanoviridae	45–120	-	*Montastraea annularis*	Metagenome	[44]
Herpes-like viral	120	-	*Montastraea annularis*	Transmission Electron Microscopy	[44]
Dicorna-like virus	30	-	*Acropora tenuis, Fungia fun-gites, Galaxea fascicularis, Pocillopora damicornis*	PCR-Based Assay	[146]
Bavuloviridae, Herpesviridae, Polydnaviridae, Retroviridae, Myoviridae	45–400	-	*Pocillopora* spp.	Metagenome	[147]
Myoviridae, siphoviridae, Mimiviridae, Baculoviridae	60–400	-	*Siderastrea siderea*	Metagenome	[148]

Wood-Charlson et al. [46] have discovered Phycodnaviridae, Marnaviridae and Alvernaviridae, which were the best characterized group of algal viruses, via metagenomic studies [149,150]. Nanoviridae and Geminiviridae are common viruses that have been found in almost every coral-associated virus study. These viruses are usually linked with sewage, which may highlight the connection between certain types of virus and environmental degradation [44]. Therefore, the abundance of viruses in corals is proportional to the concentration of local inorganic nutrients and human population centers [135,151]. According to Futch et al. [148], certain "human-specific" viruses such as Adenovirus have also been shown to exist in coral surface mucus layer, possibly due to human pollution, including sources of fecal pollution such as boats and contaminated groundwater associated with septic systems and injection wells.

According to the metagenome studies conducted by Wang et al. [152], the top five viral families that have been found in *Siderastrea sidereal* coral were Myoviridae (25.98%), Siphoviridae (9.26%), Mimiviridae (7.89%), Baculoviridae (7.61%) and Poxviridae (5.81%) whereas the overall mean number of viral abundances was about 1.14%. Moreover, similar studies have discovered thermal anomalies related to microbiome shifts, and the order Caudovirales has been found to be in high proportion in certain samplings in warmer months (July and August). Caudovirales have been found consistently in coral virome [28,46]. The member of the family known as Poxviridae is often found in marine coral viromes [140] that infect marine invertebrates yet has been found in coral, *Siderastrea sidereal,* where thermal stress increased and coral health decreased [152].

A viral outbreak in both coral species, *Acropora aspera* and *Acropora millepora,* was caused by the Herpesvirales order which has been commonly identified in these coral species. These herpesvirus-like VLPs are composed of an enveloped and circular capsid that hosts on coral epidermal and gastro-dermal cells. Moreover, the second most common abundant eukaryotic virus annotations were the NCLDVs (Nucleocytoplasmic Large DNA Viruses) including, Phycodnaviridae, Mimiviridae, Poxviridae, Iridoviridae, Marseillevirus and Ascoviridae. The phages that were found to dominate in the *Acropora aspera* virome were Sipho- and Myoviridae [43].

According to Cardenas et al. [141], the viral community composition in Red Sea corals detected via metagenomic study included 97 viral families, which were found across meta-transcriptomics. The most abundant viral families were Siphoviridae, Mimiviridae and Retroviridae (dsDNA viral families) found in eukaryotes and bacteria, whereas ss-RNA viral families such as Qinviridae, Nyamiviridae and Solinviviridae were present in meta-transcriptome studies. Parvoviridae was found to be highly abundant in stress-tolerant coral virome with massive growth in *Acanthastrea echinata, Diploastrea heliopora, Fungia* sp., and *Plerogyra sinuosa.* Dicistroviridae was observed in the outgroup samples of *Millepora platyphylla, Xenia* sp., and *Stylophora pistillata.* Similar studies have revealed the relative abundances of *Picobirnaviridae* and *Siphoviridae* that were most accounted for in viromes of *Galaxea fascicularis, Mycedium elephantotus,* and *Pachyseris speciosa* when compared to *Acropora cutherea, Pocillopora verrucosa,* and *Stylophora pistillata* [141].

5. Coral Fish Viruses

Coral disease outbreaks are the main cause of coral death and subsequent coral reef degradation [153]. It is reported that corals are susceptible to viral infections and the organisms involved mainly contribute to the HGT (Horizontal Gene Transfer) of viral particles to coral, for instance, fish, invertebrates and macroalgae [126]. Coral feeding fish (e.g., parrotfish and butterfly fish) and invertebrates [154–156] are another reasonable mechanism for the introduction of viruses into individual coral communities. According to Rotjan and Lewis [154], coral tissue mortality can be classified into two categories, where mortality of coral tissue occurred due to parrotfish grazing scars over the colony surface or mortality from unknown causes. Evidence provided by Bettarel et al. [157] that the grazing activity of the specific corallivorous gastropod *Drupella rugosa* damaged coral polyps were the result of predation on coral microbial associates is still obscure. Similar studies have revealed the presence of dsDNA virus from the coral *Acropora Formosa* which mainly belongs to uncultured Mediterranean phage uvMed, while others were Myoviridae, Podoviridae, Siphoviridae and Microviridae. The coral predators were suspected to assist in disease transmission either by acting as vectors or stressors on coral microbiota, while the presence of *Drupella rugosa* has corresponded to disease such as white syndrome, skeletal eroding band disease and black band disease [157].

Corallivorous fish act as vectors for disease transmission [158]. Aeby and Santavy [158] have determined that the coral-feeding butterflyfish, *Chaetodon capistratus* was involved in the colony transfer of black band disease. In aquaria, the presence of *Chaetodon acapistratus* increased the rate of spread from infected *Montastraea faveolata* to uninfected fragments. Both protected corals that were exposed to fish predation suffered from black band disease. Therefore, oral transmission of pathogens directly and/or via indirect fecal transmission may spread from colony to colony. This kind of direct feeding may actually be beneficial because it can reduce the extent and progression of the disease [158].

Thurber et al. [142] proposed that coral related viruses known as herpes-like viral were found to shift in response to abiotic factors when exposed to decreasing pH, elevated nutrients and thermal stress. Similar studies suggest that coral-related viruses target hosts such as protists, fungi, plants and metazoans. It is also found that invertebrates infecting viruses were mostly Baculoviruses and Polydnaviruses that predominantly infect the arthropods [159,160].

According to the study by Cherif [161], the molecular detection of Lymphocystis Disease Viral partial genome (LCDV-Sa) in Tunisian gilthead sea bream (*Sparus aurata*) caused fatal, chronic, rare and slowly developing disease which affects more than 150 marine and freshwater fish [162–164]. The etiological agent of LCD (Lymphocystis Disease Virus) belongs to genus *Lymphocystivirus,* family Iridoviridae [165,166]. *Lymphocystivirus* have been known to affect marine species such as *Holacanthus* spp., wrasses *Halichoeres* spp., grunts *Haemulon* spp., pinfish *Lagodon rhomboides*, puffers *Canthigaster* and *Sphoeroides* spp., porcupine fish *Diodon hystrix*, and many more, as stated by Stoskopf., [167]. Another study revealed LCDV strains from clownfish, *Amphiprion percula* in Indonesia. Lymphocystis in Atlantic croaker Micropogonias undulates and sand seatrout *Cynoscion arenarius* have been reported in Mississippi estuaries from 1966–1969 [168]. According to Bowden et al. [169], lymphocystis was observed in red drum *Sciaenops ocellatus* and other species in the Gulf of Mexico. In addition, a rhabdoviral infection was first described in 1983 in gray angelfish (*Pomacanthus arcuatus*) and French angelfish (*Pomacanthus paru*) collected from the Florida Keys. Table 3 below shows the reef-associated fish that have been infected with the virus.

Table 3. Lymphocystis disease virus (LCDV), detection method and its host in a coral ecosystem.

Virus Type	Host	Size (nm in Diameter)	Detection Methods	References
LCDV-Clownfish-Indonesia	*Amphiprion percula*	120–350	Polymerase Chain Reaction, LAMP	[170]
LCDV-Gilthead Sea Bream-Spain	*Sparus aurata*	120–350	Next-Generation Sequencing	[171]
LCDV-Paradise Fish-China	*Macropodus opercularis*	120–350	Electron microscopy, Polymerase Chain Reaction	[162]
LCDV-Sea Bream-Israel	*Sparus aurata*	120–350	Polymerase Chain Reaction	[172]
LCDV-Flounder-China	*Paralichthys olivaceus*	120–350	Whole-genome Shotgun Sequencing	[173]
LCDV-Largemouth Bass-USA	*Micropterus salmoides*	120–350	Nested Polymerase Chain Reaction	[174]
LCDV-Flounder-NorthSea	*Platichthys flesus*	120–350	Polymerase Chain Reaction	[165]
LCDV-Sa	*Sparus aurata*	120–350	Nested Polymerase Chain Reaction	[161]

6. Marine Sponge Viruses

Marine sponges (phylum *Porifera*) are metazoans and have been distributed all over the world in the aquatic environment since 600 million years ago [175,176]. Marine sponges are a rich source of biotechnologically potential compounds [177,178]. There is a lack of information regarding the role of sponge-related virus communities and the impact of the virus on sponge holo-biont is still unclear [179–183]. Sponge microbiome composition is shifted by environmental factors such as climate changes [184] and host sponge habitat [185] in nine sponge species [186,187]. Fan et al. [188] suggested that marine sponge symbionts lose their metabolic functional potential during the early stages of heat stress and hence destabilize the sponge holo-biont before visual signs of stress occur in the host animal. Global climate change has had a significant effect on the associated microbial communities. For example, the rise of temperature in the ocean controls the link between viruses and the cells they infect, where the growth rate of prokaryotes increases, the length of lytic cycle decreases and burst size increases, resulting in higher virus production rate [189–191]. Thus, changes in climate have direct and indirect effects on marine viruses, including impact of cascade on biogeochemical cycles, food webs and the metabolic balance of the ocean [191]. The main focus of increased studies in sponges is due to their abundance as a high source of biologically active secondary metabolites and a focal point for various aspects of research in organism origin and evolution [179,192,193].

Despite the fact that the diversity and importance of viruses in sponge-associated microorganisms are still largely unknown, a virus associated with the Lake Baikal sponge (*Lubomirskia baikalensis*) was part of the first study using cyanophage related marker gene (*g20* gene) [194] and *g23* gene for the detection of T4-like bacteriophage (Butina T.V., unpublished data). Table 4 represents marine sponge related viruses and types of method used to identify marine viruses.

Marine sponge related VLPs were investigated from the Great Barrier Reef (*Carteriospongia foliascens, Stylissa carteri, Xestospongia* sp., *Lamellodysidea herbacea, Cymbastela marshae, Cinachyrella schulzei, Pipestela candelabra* and *Echinochalina isaaci*) and Red Sea (*Xestospongia testudinaria, Amphimedon ochracea, Hyrtios erectus, Crella* (Grayela) *cyathophora,* and *Mycale* sp.) [195]. Pascelli et al. [195] revealed various types of marine sponge-associated viruses via transmission electron microscopy (TEM). Almost fifty VLP morphotypes were found in sponge tissues and mucus or surface biofilm where the viruses possessed an icosahedral symmetry morphology with diameter ranging between 60–205 nm. Moreover, the same author confirmed the presence of bacteriophage in sponge species assigned to three Caudovirales families (based on capsid symmetry and tail shape). Virus families found from marine sponge tissue and mucus in the Great Barrier Reef were Podoviridae and Siphoviridae, including Myoviridae and Inoviridae from marine sponge meso-hyl, sponge tissues and sponge mucus in the Red Sea [195].

Fan et al. [196] observed the presence of cyanophage in high abundances in sponge *Stylissa* sp. 445, suggesting that the cyanobacteria may have a lysogenic relationship with their host [197]. A few double-stranded DNA (dsDNA) viruses were abundant in a diseased branch of endemic sponge *Lubomirskia baikalensis*, for instance, Siphoviridae, Myoviridae, Phycodnaviridae, Poxviridae, and Mimiviridae; while Podoviridae, Iridoviridae and Herpesviridae can be found in healthier *Lubomirskia baikalensis* sponge [198]. The order Caudovirales with tailed bacteriophages influenced all datasets, yet the distribution of these taxa varied between holo-bionts. Phycodnaviridae, Poxiviridae, Mimiviridae, and Herpesviridae dominated all of the viromic datasets which comprised almost 98% of reads, with other unassigned viruses in Baikal sponge. Podoviridae viruses were more abundant in *Lubomirskia baikalensis* (Sv3h) healthy sponge and coral *Acropora millepora*. The presence of Herpesviridae comprised approximately 95–98% of metaviromic dataset in *Rhopaloeides odorabile* [198].

Batista et al. [199] recorded sponge Darwinella sp. and *Dysidea etheria,* with the highest abundances of Myoviridae virus detected from Arrarial do Cabo Bay site, South-Eastern Brazil, at low, upwelling and high anthropogenic influence. Laffy et al. [200] demonstrated that most dominant sponge-associated viruses detected were DNA-containing viruses, order Caudovirales. Butina et al. [201] presented virome datasets on *Baikalospongia bacillifera* species sponge in nine types of dsDNA virus families (Myoviridae, Phycodnaviridae, Siphoviridae, Poxviridae, Podoviridae, Mimiviridae, Herpesviridae, Baculoviridae and Inridoviridae) which comprises more than 70% of the identified virome sequences. Potapov et al. [202] reported that 37 nucleotide sequences of *g23* gene fragment (accession number MH576490-MH576574) were found in sponge from Lake Baikal. The data produced by Patapov et al. [202] revealed seven sequences included in Far T4 group that contains phages (*Escherichia* phage 121Q and RM378).

RNA viruses have been discovered in marine sponges, coastal water [203,204], benthic sediments [205], invertebrates [206] and vertebrates [207]. The dsRNA virus genome had been detected through advanced technology, for example, fragmented and primer-ligated dsRNA sequencing (FLDS), which obtained the full-length of the sequence of dsRNA [208–210]. Waldron et al. [211] revealed the presence of RNA virus families such as Narnaviridae, Dicistroviridae, Partitiviridae, Picobirnviridae, Picornaviridae, Tombusviridae, Nodaviridae and Herpesviridae via meta-transcriptome-FLDS analysis. Similarly, Urayama et al. [212] revealed the presence of RNA virus (Dicistroviridae) in the sponge, *Hymeniacidon* sp. Thus, they have acquired sequence encoding *RdRp* gene sequence, about 253 (2014) and 233 (2015), from Tokyo Bay which are related to nine dsRNA virus families. Of these sequences, about 78 were obtained from the sponge of *Hymeniacidon* sp. that were classified as an unassigned RNA family.

Table 4. Marine sponge associated viral communities and detection methods used to identify marine viruses.

Detection Method	Viruses	Sponge Species	References
Transmission Electron Microscopy	Podoviridae, Siphoviridae, Inoviridae, Myoviridae	*Carteriospongia foliascens, Stylissa carteri, Xestospongia* sp., *Lamellodysidea herbacea, Xestospongia testudinaria, Mycale* sp.	[195]
Metagenome	Siphoviridae, Myoviridae, Podoviridae, Phycodnaviridae, Poxviridae, Mimiviridae	*Lubomirskia baikalensis, Acropora millepora*	[198]
Metagenome	Myoviridae, Podoviridae	*Dysidea etheria, Darwinella* sp.	[199]
Metagenome	Circoviridae, Inoviridae, Polydnaviridae, Myoviridae, Siphoviridae	*Rhopaloeides odorabile*	[200]
Metavirome	Myoviridae, Phycodnaviridae, Poxviridae, Podoviridae, Mimiviridae, Herpesviridae, Baculoviridae	*Baikalospongia bacilifera*	[201]
Metagenome	*Herpes-like virus*	*Halichondria panicea*	[211]
Metagenome	Dicistoviridae	*Hymeniaciadon* sp.	[212]

7. Detection Methods for Marine Viruses in Coral Ecosystem

Diverse approaches have been used to identify and describe viruses in various organisms including transmission electron microscopy (TEM) as described in Tables 2–4 [129,213], PCR-based analyses [214], DNA in situ hybridization [215], immune-histochemistry [216], flow cytometry [217], next generation sequencing (NGS) [218] and metagenomic analyses [138,142].

Weynberg et al. [130] investigated coral associated viruses via metagenomic analysis where studies clarified that the method is a relatively promising tool for characterizing coral related viral communities. Molecular detection is frequently used by researchers in coral ecosystem associated marine viruses, which includes the viral metagenomic method and transmission electron microscopy. Sequencing of environmental DNA samples (metagenome) and sequence of other plankton viruses have revealed an unanticipated abundance of large DNA viruses associated with the marine environment [219]. Metagenomic analysis is an important tool for describing viruses, because many viral hosts are not suitable for cultivation [64] and this method does not require any of the gene markers for virus identification. Metagenomic studies have pointed out challenges and cataloged a group of worldwide viruses in corals and their symbionts [45] and this approach has widened the diversity of viral communities [130].

Although the field of coral virology is still at its early stage, some researchers have applied microscopy and genomic methods to examine the diversity and role of viruses in coral organisms. Evidence from microscopy studies has shown that virus-like particles (VLP) are present in all of the corals [29,220]. VLPs observed are likely to have been produced throughout the lytic replication phase of endogenous infections of coral animals or their microbiota [30,129]. Lawrence et al. [221] reported that the TEM approach revealed structures within corals which are marine viruses in the coral ecosystem.

Advanced molecular approaches have shown that viruses disperse in all types of environment. Sequencing of 16SRNA techniques is not recommended because they lack the gene to identify viruses, when viruses do not share common genes which significantly fit as phylogenetic markers [222]. Traditional techniques have been used to identify viruses, for example filtration, tissue culture, electron microscopy and serology. Traditional methods and advanced molecular techniques have contributed to the exploration of more viruses [222]. Electron microscopy is considered to be an expensive tool with a paucity of sensitivity. Polymerase chain reaction (PCR) only focuses on certain genes that use markers of related viruses, but PCR analyses are unable to identify complete novel viruses. In conjunction with this, metagenomic analyses are recommended to track viruses whether they are known or unknown, and they are easily detected via this method. Metagenomic study was first used in marine environmental research, where the analysis was done in San Diego [64,222], and depended upon cloning of double stranded DNA genomes. Implementing this method will discover many unidentified viruses and uncultured novel viruses which will benefit researchers in this field. Due to the difficulty of identifying some of the marine environmental viruses, especially in the ocean, a metagenomic approach would be a major breakthrough in discovering new viruses.

8. Conclusions

Marine viral communities play a crucial role in biogeochemical cycles in the ocean, indirectly and directly. These viral communities induce mortality and disease in the reef ecosystem through abiotic and biotic factors which causes the reduction in the symbiotic relationship between the coral and their surrounding hosts. This review emphasizes marine viruses from the ocean, coral-associated viruses, marine sponge and coral fish viruses in reef ecosystems, examining previous research via traditional methods to modern advanced approaches. The findings of this review have enhanced the understanding of coral-virus interactions and enriched our understanding of reef-associated virus interaction and the diversity of viral communities in marine environments.

Author Contributions: Conceptualization, S.C.Z. and S.I.; resources, L.A.; writing—original draft preparation, L.A., R.F. and E.H.S.; writing—review and editing, S.C.Z., S.I.; supervision, S.C.Z. and S.I.; funding acquisition, S.I. All authors have read and agreed to the published version of the manuscript.

Funding: This research was funded by Fundamental Research Grant Scheme, Ministry of Higher Education, Malaysia (FRGS vot. no. 59535).

Institutional Review Board Statement: Not applicable.

Informed Consent Statement: Not applicable.

Data Availability Statement: The data presented in this study are available on request from the corresponding author. The data are not publicly available due to confidentiality.

Acknowledgments: The authors would like to thank Faculty of Fisheries and Food Sciences, Universiti Malaysia Terengganu for their immense support. The authors also would like to thank the anonymous reviewers and editors for their helpful and constructive comments.

Conflicts of Interest: The authors declare no conflict of interest.

References

1. Weynberg, K.D. Viruses in Marine Ecosystems: From Open Waters to Coral Reefs. *Adv. Clin. Chem.* **2018**, *101*, 1–38. [CrossRef]
2. Mora, C.; Tittensor, D.P.; Adl, S.; Simpson, A.G.B.; Worm, B. How Many Species Are There on Earth and in the Ocean? *PLoS Biol.* **2011**, *9*, e1001127. [CrossRef] [PubMed]
3. Gazave, E.; Lapébie, P.; Ereskovsky, A.; Vacelet, J.; Renard, E.; Cárdenas, P.; Borchiellini, C. No longer Demospongiae: Homoscleromorpha formal nomination as a fourth class of Porifera. *Hydrobiologia* **2012**, *687*, 3–10. [CrossRef]
4. Wulff, J.L. Ecological interactions of marine sponges. *Can. J. Zool.* **2006**, *84*, 146–166. [CrossRef]
5. Hentschel, U.; Piel, J.; Degnan, S.; Taylor, M.W. Genomic insights into the marine sponge microbiome. *Nat. Rev. Genet.* **2012**, *10*, 641–654. [CrossRef]
6. Bell, J.J. The functional roles of marine sponges. *Estuar. Coast. Shelf Sci.* **2008**, *79*, 341–353. [CrossRef]
7. Wulff, J.L. Sponge-mediated coral reef growth and rejuvenation. *Coral Reefs* **1984**, *3*, 157–163. [CrossRef]
8. Rützler, K. Impact of crustose clionid sponges on Caribbean reef corals. *Acta Geol. Hisp.* **2002**, *37*, 61–72.
9. González-Rivero, M.; Yakob, L.; Mumby, P. The role of sponge competition on coral reef alternative steady states. *Ecol. Model.* **2011**, *222*, 1847–1853. [CrossRef]
10. Aerts, L. Sponge/coral interactions in Caribbean reefs: Analysis of overgrowth patterns in relation to species identity and cover. *Mar. Ecol. Prog. Ser.* **1998**, *175*, 241–249. [CrossRef]
11. Cole, A.J.; Pratchett, M.S.; Jones, G. Diversity and functional importance of coral-feeding fishes on tropical coral reefs. *Fish Fish.* **2008**, *9*, 286–307. [CrossRef]
12. Putnam, H.M.; Barott, K.L.; Ainsworth, T.; Gates, R.D. The Vulnerability and Resilience of Reef-Building Corals. *Curr. Biol.* **2017**, *27*, R528–R540. [CrossRef]

13. Tresguerres, M.; Barott, K.L.; Barron, M.E.; Deheyn, D.D.; Kline, D.I.; Linsmayer, L.B. Cell Biology of Reef-Building Corals: Ion Transport, Acid/Base Regulation, and Energy Metabolism. In *Acid-Base Balance and Nitrogen Excretion in Invertebrates*; Weihrauch, D., O'Donnell, M., Eds.; Springer: Cham, Switzerland, 2017. [CrossRef]
14. Shah, S.B. Coral Reef Ecosystem. In *Heavy Metals in Scleractinian Corals*; Springer: Cham, Switzerland, 2021; pp. 27–53. [CrossRef]
15. Muller-Parker, G.; D'Elia, C.F.; Cook, C.B. Interactions between Corals and Their Symbiotic Algae. In *Coral Reefs in the Anthropocene*; Birkeland, C., Ed.; Springer: Dordrecht, The Netherlands, 2015. [CrossRef]
16. Cziesielski, M.J.; Schmidt-Roach, S.; Aranda, M. The past, present, and future of coral heat stress studies. *Ecol. Evol.* **2019**, *9*, 10055–10066. [CrossRef]
17. Hoegh-Guldberg, O.; Jacob, D.; Taylor, M.W.; Guillén Bolaños, T.; Bindi, M.; Brown, S.; Camilloni, I.A.; Diedhiou, A.; Djalante, R.; Ebi, K. The human imperative of stabilizing global climate change at 1.5C. *Science* **2019**, *365*, 1–13. [CrossRef]
18. Cardini, U.; Bednarz, V.N.; Foster, R.A.; Wild, C. Benthic N2 fixation in coral reefs and the potential effects of human-induced environmental change. *Ecol. Evol.* **2014**, *4*, 1706–1727. [CrossRef]
19. de Goeij, J.M.; van Oevelen, D.; Vermeij, M.J.A.; Osinga, R.; Middelburg, J.J.; de Goeij, A.F.P.; Admiraal, W. Surviving in a marine desert: The sponge loop retains resources within coral reefs. *Science* **2013**, *342*, 108–110. [CrossRef]
20. Littman, R.; Willis, B.L.; Bourne, D.G. Metagenomic analysis of the coral holobiont during a natural bleaching event on the Great Barrier Reef. *Environ. Microbiol. Rep.* **2011**, *3*, 651–660. [CrossRef] [PubMed]
21. van Oppen, M.J.H.; Blackall, L.L. Coral microbiome dynamics, functions and design in a changing world. *Nat. Rev. Microbiol.* **2019**, *17*, 557–567. [CrossRef]
22. Bourne, D.G.; Iida, Y.; Uthicke, S.; Smith-Keune, C. Changes in coral-associated microbial communities during a bleaching event. *ISME J.* **2007**, *2*, 350–363. [CrossRef]
23. Glasl, B.; Webster, N.; Bourne, D.G. Microbial indicators as a diagnostic tool for assessing water quality and climate stress in coral reef ecosystems. *Mar. Biol.* **2017**, *164*, 91. [CrossRef]
24. Lee, S.T.M.; Davy, S.K.; Tang, S.-L.; Kench, P.S. Mucus sugar content shapes the bacterial community structure in thermally stressed Acropora muricata. *Front. Microbiol.* **2016**, *7*, 371. [CrossRef] [PubMed]
25. Vanwonterghem, I.; Webster, N.S. Coral Reef Microorganisms in a Changing Climate. *iScience* **2020**, *23*, 100972. [CrossRef] [PubMed]
26. Hughes, T.P.; Kerry, J.T.; Álvarez-Noriega, M.; Álvarez-Romero, J.G.; Anderson, K.D.; Baird, A.H.; Babcock, R.C.; Beger, M.; Bellwood, D.R.; Berkelmans, R. Global warming and recurrent mass bleaching of corals. *Nature* **2017**, *543*, 373–377. [CrossRef] [PubMed]
27. Pawlik, J.R.; Burkepile, D.E.; Thurber, R.V. A Vicious Circle? Altered Carbon and Nutrient Cycling May Explain the Low Resilience of Caribbean Coral Reefs. *Bioscience* **2016**, *66*, 470–476. [CrossRef]
28. Thurber, R.V.; Payet, J.P.; Thurber, A.R.; Correa, A.M.S. Virus–host interactions and their roles in coral reef health and disease. *Nat. Rev. Genet.* **2017**, *15*, 205–216. [CrossRef]
29. Wilson, W.H.; Chapman, D.M. Observation of virus like particles in thin sections of the plumose anemone, *Metridium senile*. *J. Mar. Biol. Assoc. U. K.* **2011**, *81*, 879–880. [CrossRef]
30. Patten, N.L.; Harrison, P.; Mitchell, J.G. Prevalence of virus-like particles within a staghorn scleractinian coral (Acropora muricata) from the Great Barrier Reef. *Coral Reefs* **2008**, *27*, 569–580. [CrossRef]
31. Leruste, A.; Bouvier, T.; Bettarel, Y. Enumerating Viruses in Coral Mucus. *Appl. Environ. Microbiol.* **2012**, *78*, 6377–6379. [CrossRef]
32. Nguyen-Kim, H.; Bettarel, Y.; Bouvier, T.; Bouvier, C.; Doan-Nhu, H.; Nguyen-Ngoc, L.; Nguyen-Thanh, T.; Tran-Quang, H.; Brune, J. Coral Mucus Is a Hot Spot for Viral Infections. *Appl. Environ. Microbiol.* **2015**, *81*, 5773–5783. [CrossRef]
33. Bettarel, Y.; Thuy, N.T.; Huy, T.Q.; Hoang, P.K.; Bouvier, T. Observation of virus-like particles in thin sections of the bleaching scleractinian coral Acropora cytherea. *J. Mar. Biol. Assoc. U. K.* **2012**, *93*, 909–912. [CrossRef]
34. Cram, J.; Parada, A.E.; Fuhrman, J.A. Dilution reveals how viral lysis and grazing shape microbial communities. *Limnol. Oceanogr.* **2016**, *61*, 889–905. [CrossRef]
35. Rohwer, F.; Thurber, R.V. Viruses manipulate the marine environment. *Nature* **2009**, *459*, 207–212. [CrossRef] [PubMed]
36. Thompson, L.; Zeng, Q.; Kelly, L.; Huang, K.H.; Singer, A.U.; Stubbe, J.; Chisholm, S.W. Phage auxiliary metabolic genes and the redirection of cyanobacterial host carbon metabolism. *Proc. Natl. Acad. Sci. USA* **2011**, *108*, E757–E764. [CrossRef]
37. Lindell, D.; Sullivan, M.B.; Johnson, Z.; Tolonen, A.; Rohwer, F.; Chisholm, S.W. Transfer of photosynthesis genes to and from Prochlorococcus viruses. *Proc. Natl. Acad. Sci. USA* **2004**, *101*, 11013–11018. [CrossRef]
38. Breitbart, M. Marine Viruses: Truth or Dare. *Annu. Rev. Mar. Sci.* **2012**, *4*, 425–448. [CrossRef] [PubMed]
39. Roossinck, M.J. Plants, viruses and the environment: Ecology and mutualism. *Virology* **2015**, *479–480*, 271–277. [CrossRef]
40. Gustavsen, J.A.; Winget, D.M.; Etian, X.; Suttle, C.A. High temporal and spatial diversity in marine RNA viruses implies that they have an important role in mortality and structuring plankton communities. *Front. Microbiol.* **2014**, *5*, 703. [CrossRef]
41. Kline, D.; Kuntz, N.; Breitbart, M.; Knowlton, N.; Rohwer, F. Role of elevated organic carbon levels and microbial activity in coral mortality. *Mar. Ecol. Prog. Ser.* **2006**, *314*, 119–125. [CrossRef]
42. Smith, J.E.; Price, N.N.; Nelson, C.E.; Haas, A.F. Coupled changes in oxygen concentration and pH caused by metabolism of benthic coral reef organisms. *Mar. Biol.* **2013**, *160*, 2437–2447. [CrossRef]

43. Correa, A.M.S.; Ainsworth, T.; Rosales, S.M.; Thurber, A.R.; Butler, C.R.; Thurber, R.L.V. Viral Outbreak in Corals Associated with an In Situ Bleaching Event: Atypical Herpes-Like Viruses and a New Megavirus Infecting *Symbiodinium*. *Front. Microbiol.* **2016**, *7*, 127. [CrossRef]
44. Soffer, N.; Brandt, M.E.; Correa, A.M.S.; Smith, T.B.; Thurber, R.V. Potential role of viruses in white plague coral disease. *ISME J.* **2014**, *8*, 271–283. [CrossRef]
45. Sweet, M.; Bythell, J. The role of viruses in coral health and disease. *J. Invertebr. Pathol.* **2017**, *147*, 136–144. [CrossRef] [PubMed]
46. Wood-Charlson, E.M.; Weynberg, K.; Suttle, C.; Roux, S.; Van Oppen, M.J.H. Metagenomic characterization of viral communities in corals: Mining biological signal from methodological noise. *Environ. Microbiol.* **2015**, *17*, 3440–3449. [CrossRef] [PubMed]
47. Suttle, C.A. Viruses in the sea. *Nature* **2005**, *437*, 356–361. [CrossRef] [PubMed]
48. Danovaro, R.; Dell'Anno, A.; Corinaldesi, C.; Magagnini, M.; Noble, R.; Tamburini, C.; Weinbauer, M. Major viral impact on the functioning of benthic deep-sea ecosystems. *Nature* **2008**, *454*, 1084–1087. [CrossRef] [PubMed]
49. Hurwitz, B.L.; Brum, J.R.; Sullivan, M.B. Depth-stratified functional and taxonomic niche specialization in the 'core' and 'flexible' Pacific Ocean Virome. *ISME J.* **2015**, *9*, 472. [CrossRef] [PubMed]
50. Spencer, R. A Marine Bacteriophage. *Nature* **1955**, *175*, 690–691. [CrossRef]
51. Weitz, J.S.; Wilhelm, S. Ocean viruses and their effects on microbial communities and biogeochemical cycles. *F1000 Biol. Rep.* **2012**, *4*, 17. [CrossRef]
52. Fuhrman, J.A. Marine viruses and their biogeochemical and ecological effects. *Nature* **1999**, *399*, 541–548. [CrossRef]
53. Kirchmann, D.L. (Ed.) *Microbial Ecology of the Oceans*; Wiley Series in Ecological and Applied Microbiology; Liss/Wiley: New York, NY, USA, 2000; 542p, ISBN 0-471-29992-8.
54. Wommack, K.E.; Colwell, R.R. Virioplankton: Viruses in Aquatic Ecosystems. *Microbiol. Mol. Biol. Rev.* **2000**, *64*, 69–114. [CrossRef]
55. Thingstad, T.F. Elements of a theory for the mechanisms controlling abundance, diversity, and biogeochemical role of lytic bacterial viruses in aquatic systems. *Limnol. Oceanogr.* **2000**, *45*, 1320–1328. [CrossRef]
56. Suttle, C.; Chan, A. Marine cyanophages infecting oceanic and coastal strains of *Synechococcus*: Abundance, morphology, cross-infectivity and growth characteristics. *Mar. Ecol. Prog. Ser.* **1993**, *92*, 99–109. [CrossRef]
57. Waterbury, J.B.; Valois, F.W. Resistance to Co-Occurring Phages Enables Marine *Synechococcus* Communities to Coexist with Cyanophages Abundant in Seawater. *Appl. Environ. Microbiol.* **1993**, *59*, 3393–3399. [CrossRef] [PubMed]
58. Sullivan, M.B.; Waterbury, J.B.; Chisholm, S. Cyanophages infecting the oceanic cyanobacterium Prochlorococcus. *Nature* **2003**, *424*, 1047–1051. [CrossRef] [PubMed]
59. Sullivan, M.B.; Coleman, M.; Weigele, P.; Rohwer, F.; Chisholm, S.W. Three Prochlorococcus Cyanophage Genomes: Signature Features and Ecological Interpretations. *PLoS Biol.* **2005**, *3*, e144. [CrossRef]
60. Culley, A.I.; Lang, A.S.; Suttle, C.A. High diversity of unknown picorna-like viruses in the sea. *Nature* **2003**, *424*, 1054–1057. [CrossRef] [PubMed]
61. Fuller, N.J.; Wilson, W.H.; Joint, I.R.; Mann, N.H. Occurrence of a Sequence in Marine Cyanophages Similar to That of T4 g20 and Its Application to PCR-Based Detection and Quantification Techniques. *Appl. Environ. Microbiol.* **1998**, *64*, 2051–2060. [CrossRef]
62. Tétart, F.; Desplats, C.; Kutateladze, M.; Monod, C.; Ackermann, H.-W.; Krisch, H.M. Phylogeny of the Major Head and Tail Genes of the Wide-Ranging T4-Type Bacteriophages. *J. Bacteriol.* **2001**, *183*, 358–366. [CrossRef] [PubMed]
63. Steward, G.; Montiel, J.L.; Azam, F. Genome size distributions indicate variability and similarities among marine viral assemblages from diverse environments. *Limnol. Oceanogr.* **2000**, *45*, 1697–1706. [CrossRef]
64. Breitbart, M.; Salamon, P.; Andresen, B.; Mahaffy, J.M.; Segall, A.M.; Mead, D.; Azam, F.; Rohwer, F. Genomic analysis of uncultured marine viral communities. *Proc. Natl. Acad. Sci. USA* **2002**, *99*, 14250–14255. [CrossRef]
65. Edwards, R.A.; Rohwer, F. Viral metagenomics. *Nat. Rev. Genet.* **2005**, *3*, 504–510. [CrossRef] [PubMed]
66. Brum, J.R.; Sullivan, M.B. Rising to the challenge: Accelerated pace of discovery transforms marine virology. *Nat. Rev. Genet.* **2015**, *13*, 147–159. [CrossRef]
67. He, T.; Jin, M.; Zhang, X. Marine Viruses. In *Virus Infection and Tumorigenesis*; Springer: Singapore, 2019; pp. 25–62.
68. Van Etten, J.L.; Dunigan, D.D.; Nagasaki, K.; Schroeder, D.C.; Grimsley, N.; Brussaard, C.P.; Nissimov, J. Phycodnaviruses (Phycodnaviridae). In *Encyclopedia of Virology*; Academic Press: Cambridge, MA, USA, 2021; pp. 687–695.
69. Breitbart, M.; Bonnain, C.; Malki, K.; Sawaya, N.A. Phage puppet masters of the marine microbial realm. *Nat. Microbiol.* **2018**, *3*, 754–766. [CrossRef] [PubMed]
70. De Corte, D.; Sintes, E.; Yokokawa, T.; Lekunberri, I.; Herndl, G.J. Large-scale distribution of microbial and viral populations in the South Atlantic Ocean. *Environ. Microbiol. Rep.* **2016**, *8*, 305–315. [CrossRef] [PubMed]
71. Yang, Y.; Motegi, C.; Yokokawa, T.; Nagata, T. Large-scale distribution patterns of virioplankton in the upper ocean. *Aquat. Microb. Ecol.* **2010**, *60*, 233–246. [CrossRef]
72. Brum, J.R.; Hurwitz, B.L.; Schofield, O.; Ducklow, H.W.; Sullivan, M.B. Seasonal time bombs: Dominant temperate viruses affect Southern Ocean microbial dynamics. *ISME J.* **2017**, *11*, 588. [CrossRef] [PubMed]
73. Roux, S.; Coordinators, T.O.; Brum, J.R.; Dutilh, B.E.; Sunagawa, S.; Duhaime, M.; Loy, A.; Poulos, B.T.; Solonenko, N.; Lara, E.; et al. Ecogenomics and potential biogeochemical impacts of globally abundant ocean viruses. *Nature* **2016**, *537*, 689–693. [CrossRef]

74. Hurwitz, B.L.; Sullivan, M.B. The Pacific Ocean Virome (POV): A Marine Viral Metagenomic Dataset and Associated Protein Clusters for Quantitative Viral Ecology. *PLoS ONE* **2013**, *8*, e57355. [CrossRef]
75. Duarte, C.M. Seafaring in the 21St Century: The Malaspina 2010 Circumnavigation Expedition. *Limnol. Oceanogr. Bull.* **2015**, *24*, 11–14. [CrossRef]
76. Walker, P.J.; Siddell, S.G.; Lefkowitz, E.J.; Mushegian, A.R.; Adriaenssens, E.M.; Dempsey, D.M.; Dutilh, B.E.; Harrach, B.; Harrison, R.L.; Hendrickson, R.C.; et al. Changes to virus taxonomy and the Statutes ratified by the International Committee on Taxonomy of Viruses (2020). *Arch. Virol.* **2020**, *165*, 2737–2748. [CrossRef]
77. Suttle, C.; Chan, A.M.; Cottrell, M.T. Infection of phytoplankton by viruses and reduction of primary productivity. *Nature* **1990**, *347*, 467–469. [CrossRef]
78. King, A.; Adams, M.; Carstens, E.; Lefkowitz, E. *Virus Taxonomy: Ninth Report of the International Committee on Taxonomy of Viruses*; Elsevier: Amsterdam, The Netherlands, 2012.
79. Paul, J.H.; Sullivan, M.B. Marine phage genomics: What have we learned? *Curr. Opin. Biotechnol.* **2005**, *16*, 299–307. [CrossRef] [PubMed]
80. Wichels, A.; Biel, S.S.; Gelderblom, H.R.; Brinkhoff, T.; Muyzer, G.; Schütt, C. Bacteriophage Diversity in the North Sea. *Appl. Environ. Microbiol.* **1998**, *64*, 4128–4133. [CrossRef] [PubMed]
81. Wilson, W.H.; Joint, I.R.; Carr, N.G.; Mann, N.H. Isolation and Molecular Characterization of Five Marine Cyanophages Propagated on *Synechococcus* sp. Strain WH7803. *Appl. Environ. Microbiol.* **1993**, *59*, 3736–3743. [CrossRef] [PubMed]
82. Maness, H.T.; Nollens, H.H.; Jensen, E.D.; Goldstein, T.; LaMere, S.; Childress, A.; Sykes, J.; Leger, J.S.; Lacave, G.; Latson, F.E.; et al. Phylogenetic analysis of marine mammal herpesviruses. *Vet. Microbiol.* **2011**, *149*, 23–29. [CrossRef]
83. Subbiah, J. Marine Viruses. In *Springer Handbook of Marine Biotechnology*; Kim, S.-K., Ed.; Springer: Berlin/Heidelberg, Germany, 2015; pp. 35–49.
84. Gangnonngiw, W.; Laisutisan, K.; Sriurairatana, S.; Senapin, S.; Chuchird, N.; Limsuwan, C.; Chaivisuthangkura, P.; Flegel, T.W. Monodon baculovirus (MBV) infects the freshwater prawn Macrobrachium rosenbergii cultivated in Thailand. *Virus Res.* **2010**, *148*, 24–30. [CrossRef]
85. Kivela, H.M.; Mannisto, R.H.; Kalkkinen, N.; Bamford, D.H. Purification and protein composition of PM2, the first lipid-containing bacterial virus to be isolated. *Virology* **1999**, *262*, 364–374. [CrossRef]
86. Farley, C.A. Viruses and virus-like lesions in marine molluscs. *Mar. Fish. Rev.* **1978**, *40*, 18–20.
87. Claverie, J.-M.; Grzela, R.; Lartigue, A.; Bernadac, A.; Nitsche, S.; Vacelet, J.; Ogata, H.; Abergel, C. Mimivirus and Mimiviridae: Giant viruses with an increasing number of potential hosts, including corals and sponges. *J. Invertebr. Pathol.* **2009**, *101*, 172–180. [CrossRef]
88. Chakrabarty, U.; Dutta, S.; Mallik, A.; Mondal, D.; Mandal, N. Identification and characterization of microsatellite DNA markers in order to recognise the WSSV susceptible populations of marine giant black tiger shrimp, Penaeus monodon. *Vet. Res.* **2015**, *46*, 1–10. [CrossRef]
89. Rector, A.; Stevens, H.; Lacave, G.; Lemey, P.; Mostmans, S.; Salbany, A.; Vos, M.; Van, D.K.; Ghim, S.J.; Rehtanz, M. Genomic characterization of novel dolphin papillomaviruses provides indications for recombination within the Papillomaviridae. *Virology* **2008**, *378*, 151–161. [CrossRef]
90. Kapp, M. Viruses infecting marine brown algae. *Virus Genes* **1998**, *16*, 111–117. [CrossRef] [PubMed]
91. Abrescia, N.G.; Cockburn, J.J.; Grimes, J.M.; Sutton, G.C.; Diprose, J.M.; Butcher, S.J.; Fuller, S.D.; Martín, C.S.; Burnett, R.M.; Stuart, D.I. Insights into assembly from structural analysis of bacteriophage PRD1. *Nature* **2004**, *432*, 68–74. [CrossRef]
92. Benson, S.D.; Bamford, J.K.H.; Bamford, D.H.; Burnett, R.M. Viral evolution revealed by bacteriophage PRD1 and human adenovirus coat protein structures. *Cell* **1999**, *98*, 825–833. [CrossRef]
93. Garseth, Å.; Biering, E.; Tengs, T. Piscine myocarditis virus (PMCV) in wild Atlantic salmon Salmo salar. *Dis. Aquat. Org.* **2012**, *102*, 157–161. [CrossRef] [PubMed]
94. Johnson, P.T. Viral diseases of marine invertebrates. *Helgol. Mar. Res.* **1984**, *37*, 65–98. [CrossRef]
95. Dietzgen, R.G.; Kondo, H.; Goodin, M.M.; Kurath, G.; Vasilakis, N. The family Rhabdoviridae: Mono- and bipartite negative-sense RNA viruses with diverse genome organization and common evolutionary origins. *Virus Res.* **2017**, *227*, 158–170. [CrossRef] [PubMed]
96. Van Bressem, M.-F.; Raga, J.A. Viruses of cetaceans. In *Studies in Viral Ecology: Animal Host Systems*; Wiley-Blackwell: Hoboken, NJ, USA, 2011; Volume 2, pp. 309–332.
97. Schütze, H. Coronaviruses in Aquatic Organisms. *Aquac. Virol.* **2016**, 327–335. [CrossRef]
98. Woo, P.C.Y.; Lau, S.K.P.; Lam, C.S.F.; Tsang, A.K.L.; Hui, S.-W.; Fan, R.Y.Y.; Martelli, P.; Yuen, K.-Y. Discovery of a Novel Bottlenose Dolphin Coronavirus Reveals a Distinct Species of Marine Mammal Coronavirus in Gammacoronavirus. *J. Virol.* **2014**, *88*, 1318–1331. [CrossRef]
99. Kapoor, A.; Victoria, J.; Simmonds, P.; Wang, C.; Shafer, R.W.; Nims, R.; Nielsen, O.; Delwart, E. A Highly Divergent Picornavirus in a Marine Mammal. *J. Virol.* **2008**, *82*, 311–320. [CrossRef]
100. Lang, A.S.; Culley, A.I.; Suttle, C. Genome sequence and characterization of a virus (HaRNAV) related to picorna-like viruses that infects the marine toxic bloom-forming alga Heterosigma akashiwo. *Virology* **2004**, *320*, 206–217. [CrossRef] [PubMed]
101. Lotz, J.M.; Overstreet, R.M.; Grimes, D.J. Aquaculture and Animal Pathogens in the Marine Environment with Emphasis on Marine Shrimp Viruses. In *Oceans and Health: Pathogens in the Marine Environment*; Springer: Boston, MA, USA, 2006; pp. 431–451.

102. Schaffer, F.L.; Bachrach, H.L.; Brown, F.; Gillespie, J.H.; Burroughs, N.; Madin, S.H.; Madeley, R.; Povey, C.; Scott, F.; Smith, A.W.; et al. Caliciviridae. *Intervirology* **1980**, *14*, 1–6. [CrossRef]
103. Greninger, A.L.; DeRisi, J.L. Draft Genome Sequences of Leviviridae RNA Phages EC and MB Recovered from San Francisco Wastewater. *Genome Announc.* **2015**, *3*, e00652-15. [CrossRef] [PubMed]
104. Bryson, S.J.; Thurber, A.R.; Correa, A.M.S.; Orphan, V.; Thurber, R.V. A novel sister clade to the enterobacteria microviruses (family Microviridae) identified in methane seep sediments. *Environ. Microbiol.* **2015**, *17*, 3708–3721. [CrossRef]
105. Nishizawa, T.; Mori, K.; Furuhashi, M.; Nakai, T.; Furusawa, I.; Muroga, K. Comparison of the coat protein genes of five fish noda-viruses, the causative agents of viral nervous necrosis in marine fish. *J. Gen. Virol.* **1995**, *63*, 1563–1569. [CrossRef]
106. Cipriano, R.C. *Infectious Salmon Anemia Virus*; Alphascript Publishing: Lewisburg, PA, USA, 2002.
107. Bonami, J.R.; Lightner, D.V.; Redman, R.M.; Poulos, B.T. Partial characterization of a togavirus (LOVV) associated with histopathological changes of the lymphoid organ of penaeid shrimp. *Dis. Aquat. Org.* **1992**, *14*, 145–152. [CrossRef]
108. Attoui, H.; Jaafar, F.M.; Belhouchet, M.; De Micco, P.; De Lamballerie, X.; Brussaard, C.P.D. Micromonas pusilla reovirus: A new member of the family Reoviridae assigned to a novel proposed genus (Mimoreovirus). *J. Gen. Virol.* **2006**, *87*, 1375–1383. [CrossRef] [PubMed]
109. Brussaard, C.; Noordeloos, A.; Sandaa, R.-A.; Heldal, M.; Bratbak, G. Discovery of a dsRNA virus infecting the marine photosynthetic protist Micromonas pusilla. *Virology* **2004**, *319*, 280–291. [CrossRef]
110. Montanié, H.; Bossy, J.-P.; Bonami, J.-R. Morphological and genomic characterization of two reoviruses (P and W2) pathogenic for marine crustaceans; do they constitute a novel genus of the Reoviridae family? *J. Gen. Virol.* **1993**, *74*, 1555–1561. [CrossRef]
111. Hosono, N.; Suzuki, S.; Kusuda, R. Genogrouping of birnaviruses isolated from marine fish: A comparison of VP2/NS junction regions on genome segment A. *J. Fish Dis.* **1996**, *19*, 295–302. [CrossRef]
112. Munn, C.B. Viruses as pathogens of marine organisms—From bacteria to whales. *J. Mar. Biol. Assoc. U. K.* **2006**, *86*, 453–467. [CrossRef]
113. Poulos, B.T.; Lightner, D.V. Detection of infectious myonecrosis virus (IMNV) of penaeid shrimp by reverse-transcriptase polymerase chain reaction (RT-PCR). *Dis. Aquat. Organ.* **2006**, *73*, 69–72. [CrossRef] [PubMed]
114. Lang, A.S.; Rise, M.L.; Culley, A.I.; Steward, G. RNA viruses in the sea. *FEMS Microbiol. Rev.* **2009**, *33*, 295–323. [CrossRef] [PubMed]
115. Zhong, X.; Rimet, F.; Jacquet, S. Seasonal variations in PCR-DGGE fingerprinted viruses infecting phytoplankton in large and deep peri-alpine lakes. *Ecol. Res.* **2014**, *29*, 271–287. [CrossRef]
116. Wilson, W.H.; Fuller, N.J.; Joint, I.R.; Mann, N.H. Analysis of cyanophage diversity in the marine environment using denaturing gradient gel electrophoresis. In *Microbial Biosystems: New Frontier, Proceedings of the 8th International Symposium on Microbial Ecology, Halifax, NS, Canada, 9–14 August 1998*; Bell, C.R., Brylinsky, M., Johnson-Green, P., Eds.; Atlantic Canada Society for Microbial Ecology: Kentville, NS, Canada, 2000; pp. 565–570.
117. Frederickson, C.M.; Short, S.M.; Suttle, C.A. The Physical Environment Affects Cyanophage Communities in British Columbia Inlets. *Microb. Ecol.* **2003**, *46*, 348–357. [CrossRef]
118. Sandaa, R.-A.; Larsen, A. Seasonal variations in virus-host populations in Norwegian coastal waters: Focusing on the cyanophage community infecting marine *Synechococcus* spp. *Appl. Environ. Microbiol.* **2006**, *72*, 4610–4618. [CrossRef] [PubMed]
119. Short, S.M.; Suttle, C.A. Sequence Analysis of Marine Virus Communities Reveals that Groups of Related Algal Viruses Are Widely Distributed in Nature. *Appl. Environ. Microbiol.* **2002**, *68*, 1290–1296. [CrossRef]
120. Short, S.; Suttle, C. Temporal dynamics of natural communities of marine algal viruses and eukaryotes. *Aquat. Microb. Ecol.* **2003**, *32*, 107–119. [CrossRef]
121. Schroeder, D.C.; Oke, J.; Hall, M.; Malin, G.; Wilson, W.H. Virus succession observed during an Emiliana huxleyi bloom. *Appl. Environ. Microbiol.* **2003**, *69*, 2484–2490. [CrossRef]
122. Martínez-Martínez, J.; Schroeder, D.C.; Larsen, A.; Bratbak, G.; Wilson, W.H. Molecular dynamics of Emiliania huxleyi and cooccurring viruses during two separate mesocosm studies. *Appl. Environ. Microbiol.* **2007**, *73*, 554–562. [CrossRef]
123. Parvathi, A.; Zhong, X.; Jacquet, S. Dynamics of various viral groups infecting autotrophic plankton in Lake Geneva. *Adv. Oceanogr. Limnol.* **2012**, *3*, 171–191. [CrossRef]
124. Luo, E.; Eppley, J.M.; Romano, A.E.; Mende, D.R.; Delong, E.F. Double-stranded DNA virioplankton dynamics and reproductive strategies in the oligotrophic open ocean water column. *ISME J.* **2020**, *14*, 1304–1315. [CrossRef] [PubMed]
125. Brüssow, H.; Hendrix, R.W. Phage Genomics: Small Is Beautiful. *Cell* **2002**, *108*, 13–16. [CrossRef]
126. Thurber, R.L.V.; Correa, A.M. Viruses of reef-building scleractinian corals. *J. Exp. Mar. Biol. Ecol.* **2011**, *408*, 102–113. [CrossRef]
127. Paul, J.H.; Rose, J.B.; Jiang, S.C.; Kellogg, C.A.; Dickson, L. Distribution of viral abundance in the reef environment of Key Largo, Florida. *Appl. Environ. Microbiol.* **1993**, *59*, 718–724. [CrossRef]
128. Wilson, W.H.; Dale, A.L.; Davy, J.E.; Davy, S. An enemy within? Observations of virus-like particles in reef corals. *Coral Reefs* **2004**, *24*, 145–148. [CrossRef]
129. Davy, J.; Patten, N. Morphological diversity of virus-like particles within the surface microlayer of scleractinian corals. *Aquat. Microb. Ecol.* **2007**, *47*, 37–44. [CrossRef]
130. Weynberg, K.D.; Wood-Charlson, E.M.; Suttle, C.A.; Van Oppen, M.J.H. Generating viral metagenomes from the coral holobiont. *Front. Microbiol.* **2014**, *5*, 206. [CrossRef]

131. Correa, A.M.S.; Welsh, R.M.; Thurber, R.L.V. Unique nucleocytoplasmic dsDNA and +ssRNA viruses are associated with the dinoflagellate endosymbionts of corals. *ISME J.* **2012**, *7*, 13–27. [CrossRef]

132. Buerger, P.; Van Oppen, M.J. Viruses in corals: Hidden drivers of coral bleaching and disease? *Microbiol. Aust.* **2018**, *39*, 9. [CrossRef]

133. Lohr, J.; Munn, C.B.; Wilson, W.H. Characterization of a Latent Virus-Like Infection of Symbiotic Zooxanthellae. *Appl. Environ. Microbiol.* **2007**, *73*, 2976–2981. [CrossRef] [PubMed]

134. Cmfauquet, C.; Fargette, D. International Committee on Taxonomy of Viruses and the 3142 unassigned species. *Virol. J.* **2005**, *2*, 1–10. [CrossRef]

135. Wegley, L.; Edwards, R.; Rodriguez-Brito, B.; Liu, H.; Rohwer, F. Metagenomic analysis of the microbial community associated with the coral Porites astreoides. *Environ. Microbiol.* **2007**, *9*, 2707–2719. [CrossRef] [PubMed]

136. Tomaru, Y.; Katanozaka, N.; Nishida, K.; Shirai, Y.; Tarutani, K.; Yamaguchi, M.; Nagasaki, K. Isolation and characterization of two distinct types of HcRNAV, a single-stranded RNA virus infecting the bivalve-killing microalga Heterocapsa circularisquama. *Aquat. Microb. Ecol.* **2004**, *34*, 207–218. [CrossRef]

137. Houldcroft, C.J.; Breuer, J. Tales from the crypt and coral reef: The successes and challenges of identifying new herpesviruses using metagenomics. *Front. Microbiol.* **2015**, *6*, 188. [CrossRef]

138. Marhaver, K.L.; Edwards, R.A.; Rohwer, F. Viral communities associated with healthy and bleaching corals. *Environ. Microbiol.* **2008**, *10*, 2277–2286. [CrossRef] [PubMed]

139. Bracht, A.J.; Brudek, R.L.; Ewing, R.Y.; Manire, C.A.; Burek, K.A.; Rosa, C.; Beckmen, K.B.; Maruniak, J.E.; Romero, C.H. Genetic identification of novel poxviruses of cetaceans and pinnipeds. *Arch. Virol.* **2005**, *151*, 423–438. [CrossRef] [PubMed]

140. Weynberg, K.D.; Laffy, P.W.; Wood-Charlson, E.M.; Turaev, D.; Rattei, T.; Webster, N.; Van Oppen, M.J. Coral-associated viral communities show high levels of diversity and host auxiliary functions. *PeerJ* **2017**, *5*, e4054. [CrossRef]

141. Cárdenas, A.; Ye, J.; Ziegler, M.; Payet, J.P.; McMinds, R.; Thurber, R.V.; Voolstra, C.R. Coral-Associated Viral Assemblages From the Central Red Sea Align With Host Species and Contribute to Holobiont Genetic Diversity. *Front. Microbiol.* **2020**, *11*, 572534. [CrossRef]

142. Thurber, R.L.V.; Barott, K.L.; Hall, D.; Liu, H.; Rodriguez-Mueller, B.; Desnues, C.; Edwards, R.A.; Haynes, M.; Angly, F.E.; Wegley, L.; et al. Metagenomic analysis indicates that stressors induce production of herpes-like viruses in the coral Porites compressa. *Proc. Natl. Acad. Sci. USA* **2008**, *105*, 18413–18418. [CrossRef] [PubMed]

143. Lawrence, S.; Davy, J.E.; Wilson, W.H.; Hoegh-Guldberg, O.; Davy, S.K. Porites white patch syndrome: Associated viruses and disease physiology. *Coral Reefs* **2014**, *34*, 249–257. [CrossRef]

144. Benites, L.F.; Silva-Lima, A.W.; Da Silva-Neto, I.D.; Salomon, P.S. Megaviridae-like particles associated with *Symbiodinium* spp. from the endemic coral Mussismilia braziliensis. *Symbiosis* **2018**, *76*, 303–311. [CrossRef]

145. Brüwer, J.D.; Agrawal, S.; Liew, Y.J.; Aranda, M.; Voolstra, C.R. Association of coral algal symbionts with a diverse viral community responsive to heat shock. *BMC Microbiol.* **2017**, *17*, 1–11. [CrossRef]

146. Montalvo-Proaño, J.; Buerger, P.; Weynberg, K.; Van Oppen, M.J.H. A PCR-Based Assay Targeting the Major Capsid Protein Gene of a Dinorna-Like ssRNA Virus That Infects Coral Photosymbionts. *Front. Microbiol.* **2017**, *8*, 1665. [CrossRef] [PubMed]

147. Messyasz, A.; Rosales, S.M.; Mueller, R.S.; Sawyer, T.; Correa, A.M.S.; Thurber, A.R.; Thurber, R.V. Coral Bleaching Phenotypes Associated With Differential Abundances of Nucleocytoplasmic Large DNA Viruses. *Front. Mar. Sci.* **2020**, *7*. [CrossRef]

148. Futch, J.C.; Griffin, D.W.; Lipp, E.K. Human enteric viruses in groundwater indicate offshore transport of human sewage to coral reefs of the Upper Florida Keys. *Environ. Microbiol.* **2010**, *12*, 964–974. [CrossRef] [PubMed]

149. Clerissi, C.; Grimsley, N.; Ogata, H.; Hingamp, P.; Poulain, J.; Desdevises, Y. Unveiling of the Diversity of Prasinoviruses (Phycodnaviridae) in Marine Samples by Using High-Throughput Sequencing Analyses of PCR-Amplified DNA Polymerase and Major Capsid Protein Genes. *Appl. Environ. Microbiol.* **2014**, *80*, 3150–3160. [CrossRef]

150. Wilson, W.H.; Van Etten, J.L.; Allen, M.J. The Phycodnaviridae: The Story of How Tiny Giants Rule the World. *Curr. Top. Microbiol. Immunol.* **2009**, *328*, 1–42. [CrossRef]

151. Dinsdale, E.A.; Pantos, O.; Smriga, S.; Edwards, R.A.; Angly, F.; Wegley, L.; Hatay, M.; Hall, D.; Brown, E.; Haynes, M.; et al. Microbial Ecology of Four Coral Atolls in the Northern Line Islands. *PLoS ONE* **2008**, *3*, e1584. [CrossRef]

152. Wang, L.; Shantz, A.A.; Payet, J.P.; Sharpton, T.J.; Foster, A.; Burkepile, D.E.; Thurber, R.V. Corals and Their Microbiomes Are Differentially Affected by Exposure to Elevated Nutrients and a Natural Thermal Anomaly. *Front. Mar. Sci.* **2018**, *5*, 5. [CrossRef]

153. Weil, E.; Smith, G.; Gil-Agudelo, D.L. Status and progress in coral reef disease research. *Dis. Aquat. Org.* **2006**, *69*, 1–7. [CrossRef]

154. Rotjan, R.D.; Lewis, S.M. Selective predation by parrotfishes on the reef coral Porites astreoides. *Mar. Ecol. Prog. Ser.* **2005**, *305*, 193–201. [CrossRef]

155. Rotjan, R.D.; Lewis, S.M. Impact of coral predators on tropical reefs. *Mar. Ecol. Prog. Ser.* **2008**, *367*, 73–91. [CrossRef]

156. Sutherland, K.P.; Porter, J.; Torres, C. Disease and immunity in Caribbean and Indo-Pacific zooxanthellate corals. *Mar. Ecol. Prog. Ser.* **2004**, *266*, 273–302. [CrossRef]

157. Bettarel, Y.; Halary, S.; Auguet, J.-C.; Mai, T.C.; Van Bui, N.; Bouvier, T.; Got, P.; Bouvier, C.; Monteil-Bouchard, S.; Christelle, D. Corallivory and the microbial debacle in two branching scleractinians. *ISME J.* **2018**, *12*, 1109–1126. [CrossRef] [PubMed]

158. Aeby, G.; Santavy, D. Factors affecting susceptibility of the coral Montastraea faveolata to black-band disease. *Mar. Ecol. Prog. Ser.* **2006**, *318*, 103–110. [CrossRef]

159. Webb, B.A.; Strand, M.R.; Dickey, S.E.; Beck, M.H.; Hilgarth, R.S.; Barney, W.E.; Kadash, K.; Kroemer, J.A.; Lindstrom, K.G.; Rattanadechakul, W.; et al. Polydnavirus genomes reflect their dual roles as mutualists and pathogens. *Virology* **2006**, *347*, 160–174. [CrossRef]

160. Herniou, E.A.; Olszewski, J.A.; O'Reilly, D.R.; Cory, J.S. Ancient Coevolution of Baculoviruses and Their Insect Hosts. *J. Virol.* **2004**, *78*, 3244–3251. [CrossRef]

161. Cherif, N.C.; Fatma, A.; Kaouther, M.; Sami, Z. Case Report: First occurrence of Lymphocystis disease virus 3 (LCDV-Sa) in Wild Marine Fish in Tunisia. *Ann. Mar. Sci.* **2020**, *4*, 024–029. [CrossRef]

162. Xu, L.; Feng, J.; Huang, Y. Identification of lymphocystis disease virus from paradise fish Macropodus opercularis (LCDVPF). *Arch. Virol.* **2014**, *159*, 2445–2449. [CrossRef]

163. Huang, X.; Huang, Y.; Xu, L.; Wei, S.; Ouyang, Z.; Feng, J.; Qin, Q. Identification and characterization of a novel lymphocystis disease virus isolate from cultured grouper in China. *J. Fish Dis.* **2014**, *38*, 379–387. [CrossRef]

164. Hossain, M.; Song, J.Y.; Kitamura, S.I.; Jung, S.J.; Oh, M.J. Phylogenetic analysis of lymphocystis disease virusfrom tropical ornamental fish species based on a major capsidprotein gene. *J. Fish Dis.* **2008**, *31*, 473–479. [CrossRef]

165. Tidona, C.; Darai, G. The Complete DNA Sequence of Lymphocystis Disease Virus. *Virology* **1997**, *230*, 207–216. [CrossRef] [PubMed]

166. Williams, T.; Barbosa-Solomieu, V.; Chinchar, V.G. A Decade of Advances in Iridovirus Research. *Adv. Appl. Microbiol.* **2005**, *65*, 173–248. [CrossRef]

167. Stoskopf, M.K. *Fish Medicine*; W. B. Saunders Co.: Philadelphia, LA, USA, 1993; 882p.

168. Christmas, J.Y.; Rouse, H.D. The occurrence of lymphocystis in Micropogon undulatus and Cynoscion arenarius from Mississippi estuaries. *Gulf Res. Rep.* **1970**, *3*, 131–154. [CrossRef]

169. Bowden, R.A.; Oestmann, D.J.; Lewis, D.H.; Frey, M.S. Lymphocystis in Red Drum. *J. Aquat. Anim. Health* **1995**, *7*, 231–235. [CrossRef]

170. Lam, C.; Khairunissa, I.; Damayanti, L.; Kurobe, T.; Teh, S.J.; Pfahl, H.; Rapi, S.; Janetski, N.; Baxa, D.V. Detection of a new strain of lymphocystis disease virus (LCDV) in captive-bred clownfish Amphiprion percula in South Sulawesi, Indonesia. *Aquac. Int.* **2020**, *28*, 2121–2137. [CrossRef]

171. López-Bueno, A.; Mavian, C.; Labella, A.M.; Castro, D.; Borrego, J.J.; Alcami, A.; Alejo, A. Concurrence of Iridovirus, Polyomavirus, and a Unique Member of a New Group of Fish Papillomaviruses in Lymphocystis Disease-Affected Gilthead Sea Bream. *J. Virol.* **2016**, *90*, 8768–8779. [CrossRef] [PubMed]

172. Kvitt, H.; Heinisch, G.; Diamant, A. Detection and phylogeny of Lymphocystivirus in sea bream Sparus aurata based on the DNA polymerase gene and major capsid protein sequences. *Aquaculture* **2008**, *275*, 58–63. [CrossRef]

173. Zhang, Q.-Y.; Xiao, F.; Xie, J.; Li, Z.-Q.; Gui, J.-F. Complete Genome Sequence of Lymphocystis Disease Virus Isolated from China. *J. Virol.* **2004**, *78*, 6982–6994. [CrossRef]

174. Hanson, L.A.; Rudis, M.R.; Vasquez-Lee, M.; Montgomery, R.D. A broadly applicable method to characterize large DNA viruses and adenoviruses based on the DNA polymerase gene. *Virol. J.* **2006**, *3*, 28. [CrossRef]

175. Kravtsova, L.S.; Izhboldina, L.A.; Khanaev, I.V.; Pomazkina, G.V.; Rodionova, E.V.; Domysheva, V.M.; Sakirko, M.V.; Tomberg, I.V.; Kostornova, T.Y.; Kravchenko, O.S.; et al. Nearshore benthic blooms of filamentous green algae in Lake Baikal. *J. Great Lakes Res.* **2014**, *40*, 441–448. [CrossRef]

176. Timoshkin, O.; Samsonov, D.; Yamamuro, M.; Moore, M.; Belykh, O.; Malnik, V.; Sakirko, M.; Shirokaya, A.; Bondarenko, N.; Domysheva, V.; et al. Rapid ecological change in the coastal zone of Lake Baikal (East Siberia): Is the site of the world's greatest freshwater biodiversity in danger? *J. Great Lakes Res.* **2016**, *42*, 487–497. [CrossRef]

177. Kamalakkannan, P. Marine sponges a good source of bioactive compounds in anticancer agents. *Int. J. Pharm. Sci. Rev. Res.* **2015**, *31*, 132–135.

178. Karuppiah, V.; Li, Z. Marine Sponge Metagenomics. In *Springer Handbook of Marine Biotechnology*; Kim, S.-K., Ed.; Springer: Berlin/Heidelberg, Germany, 2015; pp. 457–473.

179. Taylor, M.W.; Radax, R.; Steger, D.; Wagner, M. Sponge associated microorganisms: Evolution, ecology and biotechnological potentials. *Microbiol. Mol. Biol. Rev.* **2007**, *71*, 295–347. [CrossRef]

180. Hellio, C.; Maréchal, J.P.; Gama, B.P.D. Natural marine products with antifouling activities. In *Advances in Marine Antifouling Coatings and Technologies*; Hellio, C., Yebra, D., Eds.; Woodhead Publishers: Sawston, UK, 2009; pp. 572–622. [CrossRef]

181. Webster, N.; Taylor, M.W. Marine sponges and their microbial symbionts: Love and other relationships. *Environ. Microbiol.* **2011**, *14*, 335–346. [CrossRef] [PubMed]

182. Pita, L.; Rix, L.; Slaby, B.M.; Franke, A.; Hentschel, U. The sponge holobiont in a changing ocean: From microbes to ecosystems. *Microbiome* **2018**, *6*, 46. [CrossRef] [PubMed]

183. Thomas, T.; Moitinho-Silva, L.; Lurgi, M.; Björk, J.R.; Easson, C.; Astudillo-García, C.; Olson, J.B.; Erwin, P.M.; López-Legentil, S.; Luter, H.; et al. Diversity, structure and convergent evolution of the global sponge microbiome. *Nat. Commun.* **2016**, *7*, 11870. [CrossRef]

184. Lesser, M.P.; Fiore, C.; Slattery, M.; Zaneveld, J. Climate change stressors destabilize the microbiome of the Caribbean barrel sponge, Xestospongia muta. *J. Exp. Mar. Biol. Ecol.* **2016**, *475*, 11–18. [CrossRef]

185. Bourne, D.G.; Morrow, K.M.; Webster, N.S. Coral Holobionts: Insights into the coral microbiome: Underpinning the health and resilience of reef ecosystems. *Annu. Rev. Microbiol.* **2016**, *70*, 317–340. [CrossRef]

186. Easson, C.G.; Matterson, K.O.; Freeman, C.J.; Archer, S.K.; Thacker, R.W. Variation in species diversity and functional traits of sponge communities near human populations in Bocas del Toro, Panama. *PeerJ* **2015**, *3*, e1385. [CrossRef] [PubMed]

187. Souza, D.T.; Genuário, D.B.; Silva, F.S.P.; Pansa, C.C.; Kavamura, V.N.; Moraes, F.C.; Melo, I.S. Analysis of bacterial composition in marine sponges reveals the influence of host phylogeny and environment. *FEMS Microbiol. Ecol.* **2017**, *93*, fiw204. [CrossRef] [PubMed]

188. Fan, L.; Liu, M.; Simister, R.; Webster, N.; Thomas, T. Marine microbial symbiosis heats up: The phylogenetic and functional response of a sponge holobiont to thermal stress. *ISME J.* **2013**, *7*, 991–1002. [CrossRef] [PubMed]

189. Proctor, L.M.; Okubo, A.; Fuhrman, J.A. Calibrating estimates of phage-induced mortality in marine bacteria: Ultrastructural studies of marine bacteriophage development from one-step growth experiments. *Microb. Ecol.* **1993**, *25*, 161–182. [CrossRef] [PubMed]

190. Hadas, H.; Einav, M.; Fishov, I.; Zaritsky, A. Bacteriophage T4 development depends on the physiology of its host *Escherichia coli*. *Microbiology* **1997**, *143*, 179–185. [CrossRef] [PubMed]

191. Danovaro, R.; Corinaldesi, C.; Dell'Anno, A.; Fuhrman, J.A.; Middelburg, J.J.; Noble, R.T.; Suttle, C.A. Marine viruses and global climate change. *FEMS Microbiol. Rev.* **2011**, *35*, 993. [CrossRef]

192. Borchiellini, C.; Manuel, M.; Alivon, E.; Boury-Esnault, N.; Vacelet, J.; Le Parco, Y. Sponge paraphyly and the origin of Metazoa. *J. Evol. Biol.* **2001**, *14*, 171–179. [CrossRef]

193. Simion, P.; Philippe, H.; Baurain, D.; Jager, M.; Richter, D.J.; Di Franco, A.; Roure, B.; Satoh, N.; Quéinnec, É.; Ereskovsky, A.; et al. A Large and Consistent Phylogenomic Dataset Supports Sponges as the Sister Group to All Other Animals. *Curr. Biol.* **2017**, *27*, 958–967. [CrossRef]

194. Butina, T.V.; Potapov, S.A.; Belykh, O.; Belikov, S. Genetic diversity of cyanophages of the myoviridae family as a constituent of the associated community of the Baikal sponge Lubomirskia baicalensis. *Russ. J. Genet.* **2015**, *51*, 313–317. [CrossRef]

195. Pascelli, C.; Laffy, P.W.; Kupresanin, M.; Ravasi, T.; Webster, N.S. Morphological characterization of virus-like particles in coral reef sponges. *PeerJ* **2018**, *6*, e5625. [CrossRef]

196. Fan, L.; Reynolds, D.; Liu, M.; Stark, M.; Kjelleberg, S.; Webster, N.; Thomas, T.; Fan, L.; Reynolds, D.; Liu, M.; et al. Functional equivalence and evolutionary convergence in complex communities of microbial sponge symbionts. *Proc. Natl. Acad. Sci. USA* **2012**, *109*, E1878–E1887. [CrossRef]

197. Paul, J.H. Prophages in marine bacteria: Dangerous molecular time bombs or the key to survival in the seas? *ISME J.* **2008**, *2*, 579–589. [CrossRef]

198. Butina, T.V.; Bukin, Y.S.; Khanaev, I.V.; Kravtsova, L.S.; Maikova, O.O.; Tupikin, A.E.; Kabilov, M.R.; Belikov, S.I. Metagenomic analysis of viral communities in diseased Baikal sponge Lubomirskia baikalensis. *Limnol. Freshw. Biol.* **2019**, 155–162. [CrossRef]

199. Batista, D.; Costa, R.; Carvalho, A.P.; Batista, W.R.; Rua, C.P.; de Oliveira, L.; Leomil, L.; Fróes, A.M.; Thompson, F.L.; Coutinho, R.; et al. Environmental conditions affect activity and associated microorganisms of marine sponges. *Mar. Environ. Res.* **2018**, *142*, 59–68. [CrossRef]

200. Laffy, P.W.; Wood-Charlson, E.M.; Turaev, D.; Weynberg, K.; Botté, E.S.; Van Oppen, M.J.H.; Webster, N.; Rattei, T. HoloVir: A Workflow for Investigating the Diversity and Function of Viruses in Invertebrate Holobionts. *Front. Microbiol.* **2016**, *7*, 822. [CrossRef] [PubMed]

201. Butina, T.V.; Khanaev, I.V.; Kravtsova, L.S.; Maikova, O.O.; Bukin, Y.S. Metavirome datasets from two endemic Baikal sponges Baikalospongia bacillifera. *Data Brief* **2020**, *29*, 105260. [CrossRef] [PubMed]

202. Potapov, S.; Belykh, O.; Krasnopeev, A.; Galachyants, A.; Podlesnaya, G.; Khanaev, I.; Tikhonova, I. Diversity and biogeography of bacteriophages in biofilms of Lake Baikal based on g23 sequences. *J. Great Lakes Res.* **2020**, *46*, 4–11. [CrossRef]

203. Steward, G.F.; Culley, A.I.; Mueller, J.A.; Wood-Charlson, E.M.; Belcaid, M.; Poisson, G. Are we missing half of the viruses in the ocean? *ISME J.* **2013**, *7*, 672–679. [CrossRef] [PubMed]

204. Culley, A.I.; Mueller, J.A.; Belcaid, M.; Wood-Charlson, E.M.; Poisson, G.; Steward, G. The Characterization of RNA Viruses in Tropical Seawater Using Targeted PCR and Metagenomics. *mBio* **2014**, *5*, e01210-14. [CrossRef]

205. Engelhardt, T.; Orsi, W.D.; Jorgensen, B.B. Viral activities and life cycles in deep subseafloor sediments. *Environ. Microbiol. Rep.* **2015**, *7*, 868–873. [CrossRef]

206. Shi, M.; Lin, X.-D.; Tian, J.-H.; Chen, L.-J.; Chen, X.; Li, C.-X.; Qin, X.-C.; Li, J.; Cao, J.-P.; Eden, J.-S.; et al. Redefining the invertebrate RNA virosphere. *Nature* **2016**, *540*, 539–543. [CrossRef] [PubMed]

207. Shi, M.; Lin, X.-D.; Chen, X.; Tian, J.-H.; Chen, L.-J.; Li, K.; Wang, W.; Eden, J.-S.; Shen, J.-J.; Liu, L.; et al. The evolutionary history of vertebrate RNA viruses. *Nature* **2018**, *556*, 197–202. [CrossRef] [PubMed]

208. Koyama, S.; Sakai, C.; Thomas, C.E.; Nunoura, T.; Urayama, S.-I. A new member of the family Totiviridae associated with arboreal ants (*Camponotus nipponicus*). *Arch. Virol.* **2016**, *161*, 2043–2045. [CrossRef] [PubMed]

209. Urayama, S.-I.; Takaki, Y.; Nunoura, T. FLDS: A Comprehensive dsRNA Sequencing Method for Intracellular RNA Virus Surveillance. *Microbes Environ.* **2016**, *31*, 33–40. [CrossRef] [PubMed]

210. Urayama, S.-I.; Takaki, Y.; Nunoura, T.; Miyamoto, N. Complete Genome Sequence of a Novel RNA Virus Identified from a Deep-Sea Animal, Osedax japonicus. *Microbes Environ.* **2018**, *33*, 446–449. [CrossRef]

211. Waldron, F.M.; Stone, G.N.; Obbard, D.J. Metagenomic sequencing suggests a diversity of RNA interference-like responses to viruses across multicellular eukaryotes. *PLoS Genet.* **2018**, *14*, e1007533. [CrossRef]

212. Urayama, S.-I.; Takaki, Y.; Hagiwara, D.; Nunoura, T. dsRNA-seq Reveals Novel RNA Virus and Virus-Like Putative Complete Genome Sequences from Hymeniacidon sp. Sponge. *Microbes Environ.* **2020**, *35*, ME19132. [CrossRef]
213. Wilson, W.H.; Francis, I.; Ryan, K.; Davy, S.K. Temperature induction of viruses in symbiotic dinoflagellates. *Aquat. Microb. Ecol.* **2001**, *25*, 99–102. [CrossRef]
214. Chang, Y.; Cesarman, E.; Pessin, M.; Lee, F.; Culpepper, J.; Knowles, D.; Moore, P. Identification of herpesvirus-like DNA sequences in AIDS-associated Kaposi's sarcoma. *Science* **1994**, *266*, 1865–1869. [CrossRef]
215. Teifke, J.P.; Löhr, C.V.; Marschang, R.E.; Osterrieder, N.; Posthaus, H. Detection of chelonid herpesvirus DNA by nonradio-active in situ hybridization in tissues from tortoises suffering from stomatitis-rhinitis complex in Europe and North America. *Vet. Pathol.* **2000**, *37*, 377–385. [CrossRef]
216. Gouda, I.; Nada, O.; Ezzat, S.; Eldaly, M.; Loffredo, C.; Taylor, C.; Abdel-Hamid, M. Immunohistochemical Detection of Hepatitis C Virus (Genotype 4) in B-cell NHL in an Egyptian Population. *Appl. Immunohistochem. Mol. Morphol.* **2010**, *18*, 29–34. [CrossRef] [PubMed]
217. Sivaraman, D.; Yeh, H.-Y.; Mulchandani, A.; Yates, M.V.; Chen, W. Use of Flow Cytometry for Rapid, Quantitative Detection of Poliovirus-Infected Cells via TAT Peptide-Delivered Molecular Beacons. *Appl. Environ. Microbiol.* **2013**, *79*, 696–700. [CrossRef] [PubMed]
218. Marston, D.A.; McElhinney, L.M.; Ellis, R.J.; Horton, D.L.; Wise, E.L.; Leech, S.L.; David, D.; De Lamballerie, X.; Fooks, A.R. Next generation sequencing of viral RNA genomes. *BMC Genom.* **2013**, *14*, 1–12. [CrossRef] [PubMed]
219. Monier, A.; Larsen, J.B.; Sandaa, R.-A.; Bratbak, G.; Claverie, J.-M.; Ogata, H. Marine mimivirus relatives are probably large algal viruses. *Virol. J.* **2008**, *5*, 12. [CrossRef] [PubMed]
220. Pollock, F.J.; Wood-Charlson, E.M.; Van Oppen, M.J.; Bourne, D.G.; Willis, B.L.; Weynberg, K. Abundance and morphology of virus-like particles associated with the coral Acropora hyacinthus differ between healthy and white syndrome-infected states. *Mar. Ecol. Prog. Ser.* **2014**, *510*, 39–43. [CrossRef]
221. Lawrence, S.; Wilson, W.H.; Davy, J.E.; Davy, S.K. Latent virus-like infections are present in a diverse range of *Symbiodinium* spp. (Dinophyta). *J. Phycol.* **2014**, *50*, 984–997. [CrossRef]
222. Mokili, J.L.; Rohwer, F.; Dutilh, B.E. Metagenomics and future perspectives in virus discovery. *Curr. Opin. Virol.* **2012**, *2*, 63–77. [CrossRef]

Journal of
*Marine Science
and Engineering*

Article

Noble Pen Shell (*Pinna nobilis*) Mortalities along the Eastern Adriatic Coast with a Study of the Spreading Velocity

Željko Mihaljević [1], Željko Pavlinec [1], Ivana Giovanna Zupičić [1], Dražen Oraić [1], Aleksandar Popijač [2], Osvin Pećar [2], Ivan Sršen [2], Miroslav Benić [3], Boris Habrun [3] and Snježana Zrnčić [1,*]

[1] Department of Pathology, Croatian Veterinary Institute, 10000 Zagreb, Croatia; miha@veinst.hr (Ž.M.); pavlinec@veinst.hr (Ž.P.); zupicic@veinst.hr (I.G.Z.); oraic@veinst.hr (D.O.)
[2] Mljet National Park, Goveđari, 20226 Mljet, Croatia; aleksandar.popijac@np-mljet.hr (A.P.); osvin.pecar@np-mljet.hr (O.P.); ivan.srsen@np-mljet.hr (I.S.)
[3] Department of Bacteriology and Parasitology, Croatian Veterinary Institute, 10000 Zagreb, Croatia; benic@veinst.hr (M.B.); habrun@veinst.hr (B.H.)
* Correspondence: zrncic@veinst.hr

Abstract: Noble pen shells (*Pinna nobilis*) along the Eastern Adriatic coast were affected by mass mortalities similarly to the populations across the Mediterranean basin. Samples of live animals and organs originating from sites on Mljet Island on the south and the Istrian peninsula on the north of the Croatian Adriatic coast were analyzed using histology and molecular techniques to detect the presence of the previously described *Haplosporidium pinnae* and *Mycobacterium* spp. as possible causes of these mortalities. To obtain more information on the pattern of the spread of the mortalities, a study was undertaken in Mljet National Park, an area with a dense population of noble pen shells. The results of the diagnostic analysis and the velocity of the spread of the mortalities showed a significant correlation between increases in water temperature and the onset of mortality. Moderate to heavy lesions of the digestive glands were observed in specimens infected with *H. pinnae*. A phylogenetic analysis of the detected *Haplosporidium pinnae* showed an identity of 99.7 to 99.8% with isolates from other Mediterranean areas, while isolated *Mycobacterium* spp. showed a higher heterogeneity among isolates across the Mediterranean. The presence of *Mycobacterium* spp. in clinically healthy animals a few months before the onset of mortality imposes the need for further clarification of its role in mortality events.

Keywords: noble pen shell; mass mortality event; Croatia; *Haplosporidium pinnae*; *Mycobacterium* spp.; spreading velocity

Citation: Mihaljević, Ž.; Pavlinec, Ž.; Zupičić, I.G.; Oraić, D.; Popijač, A.; Pećar, O.; Sršen, I.; Benić, M.; Habrun, B.; Zrnčić, S. Noble Pen Shell (*Pinna nobilis*) Mortalities along the Eastern Adriatic Coast with a Study of the Spreading Velocity. *J. Mar. Sci. Eng.* **2021**, *9*, 764. https://doi.org/10.3390/jmse9070764

Academic Editor: Valerio Matozzo

Received: 14 June 2021
Accepted: 9 July 2021
Published: 12 July 2021

Publisher's Note: MDPI stays neutral with regard to jurisdictional claims in published maps and institutional affiliations.

1. Introduction

As in many parts of the Mediterranean area, the noble pen shell (*Pinna nobilis*) is the largest bivalve (60–120 cm) and an endemic inhabitant of shallow waters along the Croatian Adriatic coastline [1]. Over the last few decades, its numbers have drastically declined [2], and the species is now protected under Annex IV of the Habitats Directive, Annex II of the Barcelona Convention, and national legislation in Croatia and most Mediterranean countries. The decline has been attributed to uncontrolled trawling [3], illegal collection for food or souvenirs by amateur divers, and devastation of their natural habitats due to anthropogenic inputs [1,4]. Some coastal waters in the Mediterranean basin that are known as natural habitats of this protected species have been designated as marine parks, such as the Mljet Island National Park in Croatia or the Parque Natural de Cabo de Gata-Nijar [5] and the Parque Nacional Marítimo-Terrestre del Archipiélago de Cabrera in Spain [6].

Unfortunately, in early autumn of 2016, mass mortalities that affected the noble pen shell at several different points along the Spanish Mediterranean coast were reported [7], indicating the presence of a haplosporidian-like parasite within the digestive gland as a possible cause of the mortalities. Shortly thereafter, severe mortalities spread to most of the

Spanish coast, Corsica in France [8], several sites in Italy [9,10], and the Aegean Sea [11,12]. Recently, they were also reported in the Adriatic Sea, affecting the populations in Albania and Croatia [13–15], as well as in Turkey [16], Tunisia, and Morocco [17]. These mass mortality outbreaks are widespread and worrying. Consequently, *P. nobilis* was included in the red list of critically endangered species [18].

Studies on the etiology described the presence of *Haplosporidium* species that parasitized the digestive epithelium. As a result of the morphology and molecular phylogeny, the new species *Haplosporidium pinnae* was designated [8]. The same parasite (morphologically and phylogenetically) was reported to be associated with inflammation and heavy lesions in the digestive tubules that were a cause of a mass mortality event (MME) in the Gulf of Taranto (Ionian Sea) in southern Italy [9]. Nonetheless, Carella et al. [10] observed strong inflammatory lesions in the connective tissue surrounding the digestive system and gonads and linked these to the presence of *Mycobacterium* spp. In the same animals, *Haplosporidium pinnae* was observed. Hence, the MMEs were attributed to co-infection of the parasite and bacterium. This statement was supported by the findings of co-infections of the same pathogens in moribund pen shells collected from MME sites in Greece [11,12]. However, some other pathogens, such as bacteria from the different genera (e.g., *Vibrio, Rhodococcus* sp., and *Mycoplasma*) and parasites of the genus *Perkinsus* [16,19–24], were also detected in noble pen shells.

One of the most important habitats of *P. nobilis* in the southern Eastern Adriatic Sea is in Mljet National Park [25], which was two lake-like inlets: The Small Lake and Big Lake. Šiletić and Peharda [26] found that the density of the individuals in this area appeared to be higher than in other parts of the Adriatic and Mediterranean areas, and that the population comprised primarily of adults and older individuals that were potentially aged 8–15 years [5,27]. To monitor the health status of the *P. nobilis* in the Adriatic Sea, we conducted an inspection of the *P. nobilis* populations, primarily in the Mljet National Park and in habitats along the Istrian peninsula, from April 2019 until May 2020.

The main objectives of this study are (1) to present the results of laboratory analyses of pen shells collected from habitats along the Istrian peninsula and the Mljet National Park; (2) to report the observations of a small-scale study carried out in the Mljet National Park to elucidate the rate of the spread of MMEs within dense populations; and (3) to implement and assess a new non-lethal method of mantle tissue sampling for diagnostic purposes.

2. Materials and Methods

2.1. Sampling Sites and Sample Collection

In this study, we collected noble pen shells from two habitats along the Eastern Adriatic coast: The Mljet National Park (MNP) and the Istrian peninsula. The sampling sites were determined according to the previously collected information on the natural habitats of *P. nobilis* (Figures 1 and 2; Table 1). Samples of live animals were collected in the MNP by scuba diving during April 2019, and they were delivered to the laboratory under cooling conditions with temperatures not exceeding 8 °C. The other samplings—both along the Istrian peninsula and in the MNP—were also carried out by scuba diving. All the samplings were carried out with the permission of the Croatian Ministry of Environmental Protection and Energy (CLASS UP/1-612-17/18-48/172; No. 517-05-1-1-18-4 of 21 December 2018 and CLASS UP/1-612-07/19-48/193; No. 517-05-1-1-19-3 of 11 September 2019).

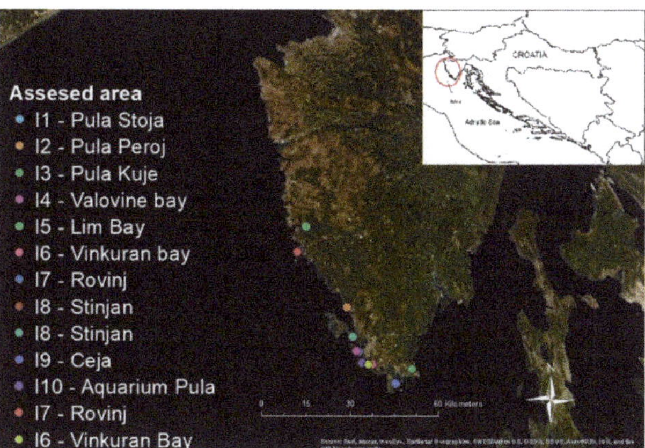

Figure 1. Sampling sites along the Istrian peninsula in the northern part of the Eastern Adriatic coast.

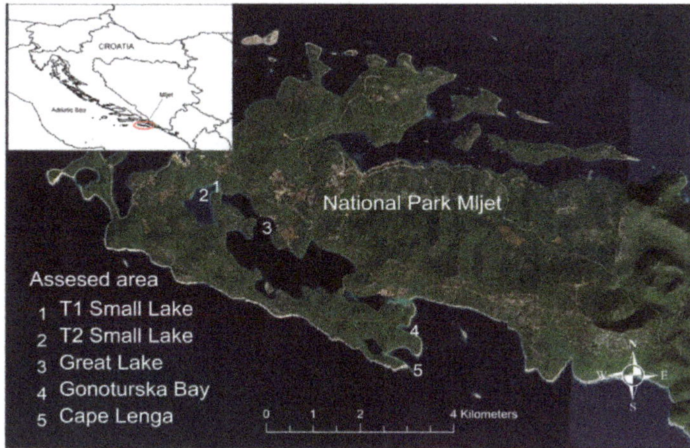

Figure 2. Study site within the Mljet National Park with five sampling points and two transect areas.

During the sampling, the appearance of pen shells in their habitat was evaluated, and portions of digestive glands, gills, mantles, gonads, and muscles were collected separately and preserved in Davidson solution [28] for histological examination and in either EtOH or RNAlater (Thermo Fisher, Waltham, MA, USA) for molecular analysis. Exceptionally, 20 mantle tissue biopsies were taken during September 2019 with a non-lethal sampling method and were preserved in EtOH.

2.2. Site and Design of the Small-Scale Study within the MNP

The study was conducted from April to November of 2019 in the MNP, which is in the south-east Adriatic (Figure 2), a highly productive and biodiverse marine ecosystem within the Natura 2000 European Network of Protected Areas (code HR5000037). Five sampling points were selected according to the most abundant pen shell populations. Two sampling points (1 and 2) were situated in the Small Lake, a lake-like inlet connected by a shallow, narrow channel, while sampling point 3 was in the Big Lake, which was connected to the open sea by a slightly deeper, wider channel. Another two sampling points in addition to these lake-like inlets were Gonoturska Bay (sampling point 4) and Cape Lenga (sampling point 5). The distance between the most external sampling point

(5) and the most internal sampling points (1 and 2) was about four kilometers. Based on the discovery of *Haplosporidium pinnae* at the outer sites in April, a field experiment was set up to determine the pattern and rate of the spread of mortality within this dense population. As it was obvious that MMEs started to spread quickly through the inner part of the MNP, we set up the study with the aim of learning more about the pattern of the spread of the disease.

Two 100 m surveillance sentinel transects—transect 1 (T1) and transect 2 (T2)—were placed in the Small Lake (Figure 2) in September 2019, perpendicularly to the shoreline and extending towards a deeper part of the inlet (Figure 3). The deepest points along the transects were 3.8 m for T1 and 7.3 m for T2. All the clinically healthy pen shells were tagged, and their positions along the transects were recorded within a 2 m corridor on each side of the transect line, covering approximately 400 m^2 along each transect. A caliper was used to determine the maximum unburied length (UL) and minimum dorsoventral height or minimum width (w), and the maximum height was determined using the formula Ht = 1.79 w + 0.5 + UL, as described by García-March and Vicente [29]. The data were recorded underwater on a plastic slate, while the depth was determined using a diving computer. Along transect T1, there were 75 individuals with the density of 20.8 per 100 m^2, while along T2, there were 106 individuals with a density of 37.83 individuals per 100 m^2. Every 2 weeks from September to December 2019, two divers clinically examined all the pen shells along both transects and noted the results of the clinical examinations and the numbers of live specimens. The temperature and the salinity of the sea were measured and noted daily.

Figure 3. Labeling healthy individuals along the transect.

2.3. Non-Lethal Sampling

As the mortality rate within the study area rose to 20% in September, biopsies of the mantle tissue from 20 randomly selected individuals along both transects were taken to assess the non-lethal sampling methodology and to obtain information on the prevalence of infections. The biopsies of the mantles were collected according to the method described by Sanna et al. [30]. Obtaining biopsies did not cause significant damage since the individuals' valves were held open with a wooden stick (diameter = 0.5 cm), which was placed in proximity (4–5 cm) to the hinge ligament, and about 50 mg of mantle tissue samples were removed using forceps. The stick was then removed, and the tissue sample was stored in a 5 mL tube. Once the diver returned to the surface, the tissue samples were preserved in EtOH and transferred to the laboratory in a refrigerated box.

2.4. Statistical Analysis

Differences in the morphometric data between transects were compared with a multi-variant regression analysis using STATA 13.1 (Stata, College Station, TX, USA). Survival curves were calculated according to the Kaplan–Meier method [31].

2.5. Laboratory Procedures

The live mollusks that were sent to the laboratory were opened by cutting the adductor muscle according to the methods described by Morton and Puljas [32]. The in situ appearance of the organs was evaluated. Samples of the gills, mantle, digestive gland, gonads, and muscles were collected from each individual for both molecular and histological purposes and were immediately preserved in ethanol or Davidson fixative, respectively [28].

2.6. Histological Analysis

Samples of the organs fixed in Davidson solution were processed for histological examination. Tissues were dehydrated with an ethanol series, cleared in xylene, embedded in paraffin, sectioned at 3 μm, and mounted on Microme EC 350-2 slides (Thermo Scientific, Waltham, MA, USA). Mounted slides were heated to 60 °C, deparaffinized in xylene, and rehydrated in an ethanol series and distilled water, followed by staining with hematoxylin and eosin (H&E, Harris).

2.7. DNA Extraction

DNA was extracted from approximately 25 mg of the gill, mantle, digestive gland, gonads, and muscle tissues, respectively, which were preserved in either EtOH or RNAlater, using a MagMAX CORE Nucleic Acid Purification Kit (Applied Biosystems, Foster City, CA, USA) according to the manufacturer's instructions on a KingFisher Duo Prime Purification System (Thermo Scientific, USA). The extracted DNA was stored at −20 °C until the analysis.

2.8. PCR for Detection of Haplosporidium spp.

For the detection of *Haplosporidium* spp., the *Haplosporidium*-specific primers described by Renault et al. [33] were used. We used 2 μL of extracted DNA and 0.5 μM of the primers HAP-F1 (5′-GTTCTTTCWTGATTCTATGMA-3′) and HAP-R2 (5′-GATGAAYAATTGCAAT CAYCT-3′). PCR reactions were performed using the GoTaq G2 Hot Start Colorless Master Mix (Promega, Madison, WI, USA) on a ProFlex PCR System (Applied Biosystems, USA) with a final volume of 20 μL. The temperature protocol involved enzyme activation at 95 °C for 2 min, followed by 40 cycles of denaturation at 94 °C for 30 s, primer annealing at 48 °C for 30 s, and elongation at 72 °C for 60 s, and the process was completed with a final elongation step at 72 °C for 5 min. The results of the PCRs were checked through electrophoresis on a QIAxcel system (Qiagen, Hilden, Germany).

2.9. PCR for Detection of Mycobacterium spp.

For the detection of *Mycobacterium* spp., we used two protocols. For the amplification of a segment of the *hsp65* gene, we used the primers Tb11 (5′-ACCAACGATGGTGTGTCCAT-3′) and Tb12 (5′-CTTGTCGAACCGCATACCCT-3′), which were first described by Telenti et al. [34]. The thermal protocol involved enzyme activation at 95 °C for 2 min, followed by 45 cycles of denaturation at 94 °C for 60 s, primer annealing at 60 °C for 60 s, and elongation at 72 °C for 60 s, and the process was completed with a final elongation step at 72 °C for 5 min. For the amplification of a segment of the mycobacterial 16S rRNA gene, we used the primers 246 (5′-AGAGTTTGATCCTGGCTCAG-3′) and 264 (5′-TGCACACAGGCCACAAGGGA-3′), which were first described by Böddinghaus et al. [35]. The thermal protocol was the same as that with the first set of primers, except for the annealing temperature, which was 55 °C. All the PCR reactions were performed using the GoTaq G2 Hot Start Colorless Master Mix (Promega) on a ProFlex PCR System (Applied Biosystems). For each reaction, we used 2 μL of extracted DNA and 0.5 μM primers with a final volume of 20 μL.

2.10. Sequencing and Phylogeny

To obtain longer sequences of the haplosporidian 18S rRNA gene for phylogenetic comparisons, for all the samples that tested positive for haplosporidian DNA, we performed PCR with the primers HPN-F1, HPN-F3, HPN-R3, and 16SB, as described by Catanese et al. [7]. All the PCR products obtained were sequenced, and the resulting sequences were aligned and assembled in continuous reads using MAFFT version 7 [36]. For a phylogenetic comparison of *Mycobacterium* spp., the PCR products obtained using 246–264 primer pairs were sequenced. Direct Sanger sequencing was performed by Macrogen Europe (Amsterdam, The Netherlands). The sequences obtained [37] were identified using BLAST [38]. The sequences were aligned with MAFFT v7.388 [37] with the default parameters. The phylogenetic analysis was performed in MEGA 10.0.5 [39]. The best model was selected based on the lowest BIC score. Maximum likelihood trees were calculated with the following settings: Five discrete gamma categories, use of all sites, SPR level 5 heuristic method with a very strong branch-swap filter, the K2+G substitution model for the haplosporidian tree, and the HKY+G+I substitution model for the *Mycobacterium* spp. tree. Phylogeny was tested with the bootstrap method with 1000 replications. The obtained tree was visualized in iTOL [40].

3. Results

During the study, the health checks of noble pen shells (*Pina nobilis*) collected on 15 sampling sites were included, five in the MNP and 10 along the Istrian coast during 2019 and the first half of 2020 (Figures 1 and 2; Table 1).

Table 1. List of sampling points and samples collected during 2019 and 2020, which were analyzed using PCR for detection of *Haplosporidium* spp. (Haplo) and *Mycobacterium* spp. (Myco) and using histology.

Date of Sampling (D/M/Y)	Name and Number of the Site	Sea Temp. °C	Sample	Results of Laboratory Analysis		
				PCR Haplo	PCR Myco	Histology
19/04/19	MNP1—Small Lake	17.2	Whole animal	0/3	2/3	neg
19/04/19	MNP5—Cape Lenga	15.9	Whole animal	2/2	0/2	Haplo +
19/04/19	MNP4—Gonoturska	15.9	Whole animal	0/1	0/1	neg
04/09/19	I1—Pula-Stoja	24.5	Organs in EtOH *	0/3	0/3	n/d
04/09/19	I2—Pula-Peroj	24.8	Organs in EtOH	0/1	0/1	n/d
04/09/19	I3—Pula-Kuje	25.2	Organs in EtOH	0/1	0/1	n/d
24/09/19	MNP1 Mljet-Small Lake	28	Mantles in EtOH	2/4	1/4	n/d
30/09/19	MNP1 Mljet-Small Lake	26	Mantles in EtOH	2/16	1/16	n/d
13/01/20	I4—Uvala Valovine	12.3	Organs in EtOH	0/1	0/1	n/d
13/01/20	I5—Limski Bay	12.8	Organs in RNAlater	0/1	0/1	n/d
18/02/20	I6—Vinkuran Bay	12.2	Organs in EtOH	0/1	0/1	n/d
19/02/20	I7—Rovinj	12.2	Organs in EtOH	0/2	0/2	n/d
20/02/20	I8—Štinjan	11.9	Organs in EtOH	0/1	0/1	n/d
19/05/20	I8—Štinjan	21.5	Organs in EtOH	2/2	2/2	n/d
19/05/20	I9—Ceja	22.7	Organs in EtOH	1/1	0/1	n/d
28/05/20	I10—Aquarium Pula	18.0	Organs in EtOH and HCHO **	0/1	1/1	n/d
28/05/20	I7—Rovinj	18.5	Organs in EtOH and HCHO	1/1	1/1	Haplo +
28/05/20	I6—Vinkuranska vala	17.8	Organs in EtOH and HCHO	1/1	0/1	Haplo +

Legend: MNP: Mljet National Park; I: Istrian peninsula; 2/2: Two positive animals out of two tested; n/d: Analysis not performed due to the absence of appropriate samples; * EtOH: Ethanol; ** HCHO: Formaldehyde.

3.1. Description of the Pen Shells Collected in Situ

During the monitoring of the sampling sites included in this study, we observed the different appearances of the present noble pen shells. Healthy individuals, individuals with disease symptoms, and even empty shells were observed (Figure 4). The external symptoms were very scarce, and the affected animals displayed weakened reactions to stimuli, such as gaping (slower closing of shells) after touching or increased drifting and obvious shrinkage of the mantle, causing the illusion of a greater space between the shells than there was. At the sites with high mortalities of pen shells, many empty shells remained in an upward position, causing the illusion that the animals were still alive. In this situation, soft tissues were never found. When the sick animals were dissected, a certain amount of liquid was observed within the shells, and the mantle was darker than usual (Figure 5).

Figure 4. Clinical inspection of the live noble pen shells in situ, where both healthy and sick animals were in an upward position. Differentiation between them was made by checking their reactions to stimuli. The healthy specimens quickly closed their shells.

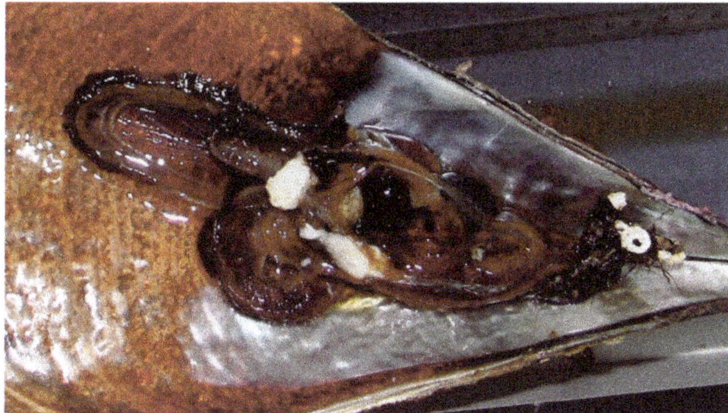

Figure 5. Darker mantle and shrunken soft tissues in a noble pen shell infected by *Haplosporidium pinnae*.

3.2. Results of the Pilot Study (Morphometrics, Environmental Conditions, and Mortality Patterns)

During the first health screening in the MNP, which was carried out in mid-April 2019, a total of six specimens were collected in the sampling points shown in Figure 2. Three of them were collected in the Small Lake (sampling points 1 and 2), and no changes in appearance were observed. However, two out of the three samples tested positive for *Mycobacterium* spp. The others were collected outside the lake-like inlets, one in Gonoturska (sampling point 4) and two around Cape Lenga (sampling point 5). In Gonoturska (sampling point 4), there were no signs of unusual behavior, and a single sample tested negative for both pathogens. Around Cape Lenga (sampling point 5), a high mortality rate of up to 80% was noticed, which was based on the presence of unresponsive specimens and empty shells. In the other four sampling points, the mortality rate was about 9%, which is considered normal. Laboratory tests revealed the positive discovery of *Haplosporidium pinnae* in both individuals from sampling point 5—with both histology (Figure 7a–d) and PCR—while both specimens tested negative for *Mycobacterium* spp. (Table 1). At the beginning of September, a mortality rate of almost 100% was observed in the Big Lake, and a rate of about 20% was observed in the Small Lake. Following the spread of mortality in the Big Lake and when setting up the pilot study, we found that all the individuals were adults, and those along T1 had a mean height of 37.1 cm (±0,65), mean width of 15.28 cm (±0.35), and calculated mean maximum height of 64.56 cm (±1.11). The mean height of the individuals along T2 was 39.19 cm (±0.56), they had a width of 14.77 cm (±0.28), and the calculated maximum height was 65.64 cm (±0.76). The detailed morphometric properties of the individuals marked along both transects did not differ significantly (Supplement 1). At the beginning of the surveillance in April, the temperature of the sea was 15 °C, and it increased over the following months, reaching 27 °C in August, and slowly decreased over the study period from 25 °C at the beginning of September to 10 °C in December. During the whole study period, the salinity had an average value of 37.98%—ranging from 35.30 to 39.20%—on the surface or 38.04%—ranging from 35.30 to 39.00%—at a depth of 2 m.

The mortality rate during April 2019 at sampling point 5 was estimated to be 80%, and in the rest of the study area, it was estimated to be about 9% based on the number of empty shells (Figure 2). During the control diving on August 1, the mortality in the Big Lake was about 20%, and at the beginning of September, 100% mortality was observed in the Great Lake, also in addition to the increased mortality in the Small Lake (Table 2). At the time of setting up the transects, along T1, there were 62 dead individuals out of a total of 137 (Mt = 45.26%), and along T2, there were 28 dead individuals out of 134 (Mt = 20.90%). Over the study period, the mortality rate increased, resulting in the final survival of one individual from the T1 and three individuals from T2.

Table 2. Number of dead noble pen shells found along the transects from the setup of the pilot study (4 September 2019) until its completion (25 November 2019).

Date	Number of Dead Noble Pen Shells found in the Small Lake	
	Transect 1 (T1)	**Transect 2 (T2)**
4 September	62	28
18 September	27	25
26 September	3	5
9 October	8	15
24 October	29	28
31 October	5	9
8 November	1	12
25 November	1	9
Total	74/75	103/106

The Kaplan–Meier survival curves indicate two mass mortality events during the study period: The first was between 4 and 18 September, when 52 animals died, and the second was between 9 and 24 October 2019, when 57 animals died (Figure 6 and Table 2). By 25 November 2019, one of the initial 75 pen shells were alive along T1, corresponding to a survival rate of 1.33%, while three of the initial 106 pen shells were alive along T2, corresponding to a survival rate of 2.83%. The overall survival rate was 2.21%.

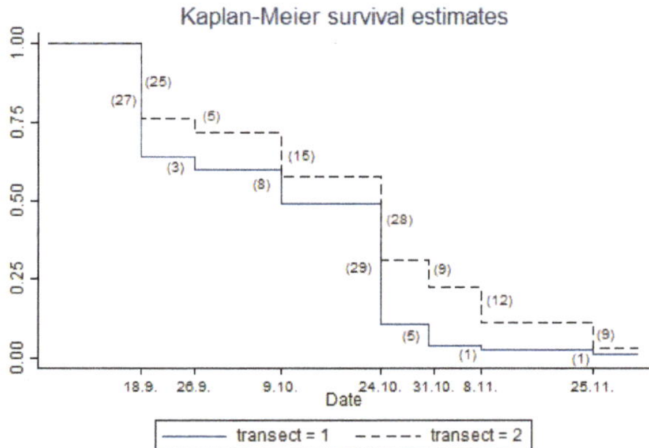

Figure 6. Kaplan–Meier survival estimates.

During the clinical checks, we observed that the mortalities affected certain individuals and spread diffusely to others in the close vicinity, affecting primarily those in the shallower area of the Small Lake and, later, those in the deeper parts.

3.3. Necropsy and Histological Findings

Six live animals were sacrificed for health evaluation, and organs from 37 additional animals were collected for molecular analysis (Table 1). Live animals were collected before the first notification of the MME, while the organs were collected prior, during, and after the events. On dissection, two of the animals collected in April showed a darker mantle and reduced soft tissues (Figure 5). The results of the histological analyses in the infected animals disclosed moderate to heavy infiltration of hemocytes into the connective tissues of the digestive gland and gonads, as well as light necrosis in the connective tissues of the gills, digestive gland, and muscle (Figure 7a). Uni-nucleated *Haplosporidium*-like parasite stages were observed in the connective tissues of the mantle (Figure 7b) and musculature (Figure 7c). Different *Haplosporidium*-like parasite stages, such as uni- or binucleate stages and plasmodia, were present in the connective tissues and epithelia of the digestive glands (Figure 7d,e). Intra-hemocytic stages of uninucleate cells were observed in the heavily infected specimens, in which sporogonia were observed in the epithelia of the digestive gland (Figure 7e,f). However, very few specimens were analyzed using histology, and sporulation was observed in the digestive gland in the sample collected in May 2020 in Vinkuranska vala in the Istrian peninsula.

3.4. Molecular Analysis for the Presence of Haplosporidium spp. and Mycobacterium spp.

Eleven out of the 43 animals tested positive for *Haplosporidium* spp. (Table 1). Since the DNA was isolated from the digestive glands, mantles, gonads, and muscles of the infected animals, mostly the digestive glands and gonads, followed by muscles and, in a few animals, mantles were positive for *Haplosporidium* spp. Here, it should be emphasized that four biopsies of mantles tested positive for *Haplosporidium* spp., as well. Furthermore, eight out of 43 samples tested positive for *Mycobacterium* spp. As in the case of *Haplosporidium* spp., two of the positive cases were found with biopsies.

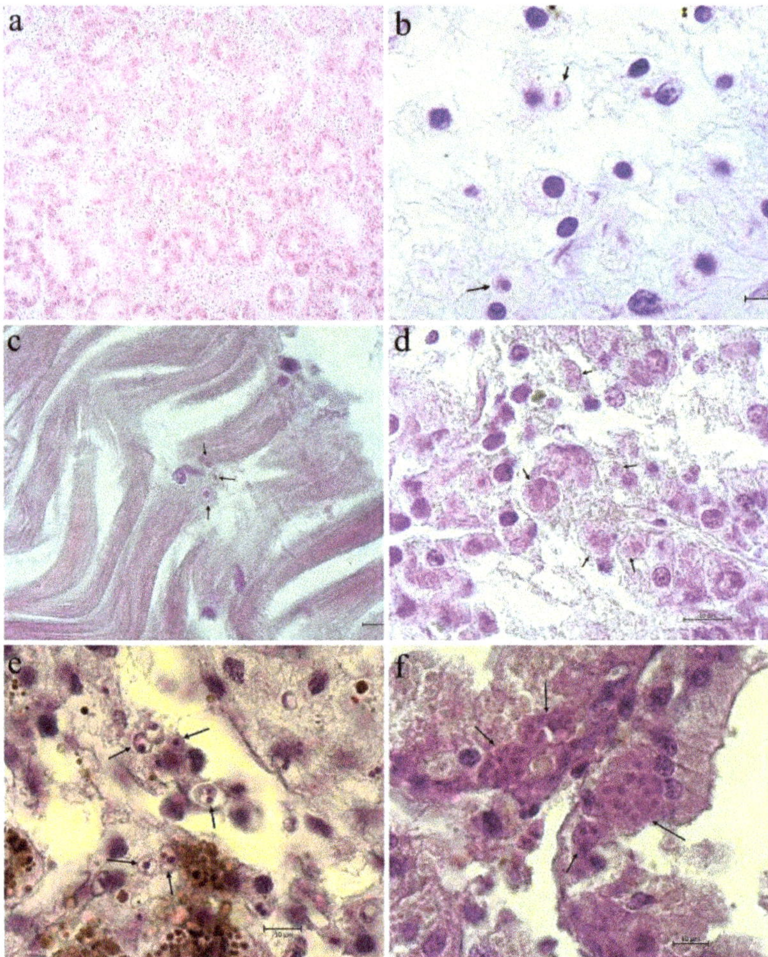

Figure 7. Histological findings in the organs of noble pen shells: Moderate infiltration of hemocytes into the connective tissue of the digestive gland (**a**); uni- and binucleate cells of *H. pinnae* in the connective tissue of the mantle (**b**); uninucleate cells of *H. pinnae* in the muscle (**c**); uninucleate cells and plasmodia in the connective tissue of the digestive gland (**d**); intra-hemocytic stages of uninucleate cells in the connective tissue of the digestive gland (**e**); sporulation stages of *H. pinnae* in the epithelium of the digestive gland with a disruption of the digestive tubules (**f**).

3.4.1. Sequencing of *Haplosporidium* spp. Isolates

All the *Haplosporidium*-positive PCR products (1451 bp) were successfully sequenced, and since the obtained sequences did not differ, only one haplotype was deposited in the GenBank (accession number MT367896). The sequences of the Croatian isolates (Figure 8) showed 99.8% identity with the Italian isolates deposited by Scarpa et al. [23] and Tiscar et al. (unpublished data) and 99.7% with the Spanish isolates [8], confirming the presence of *Haplosporidium pinnae* in all the positive noble shells.

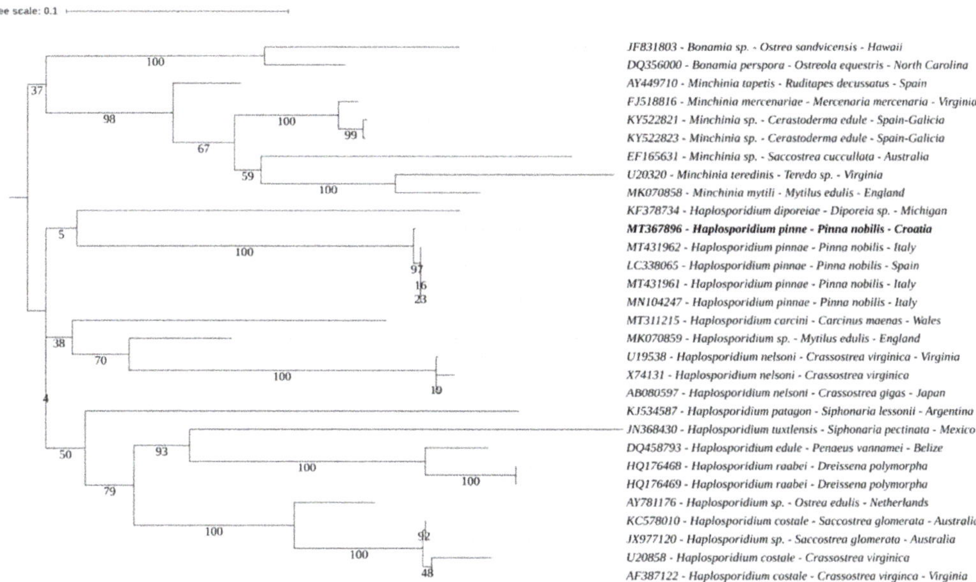

Figure 8. Maximum likelihood phylogenetic tree for haplosporidians. The numbers on the branches represent bootstrap values.

3.4.2. Sequencing of *Mycobacterium* spp. Isolates

In addition, all the PCR products that tested positive for *Mycobacterium* spp. (985 bp) were sequenced, and they were identical. For this reason, only one haplotype was deposited in the GenBank (accession number MT367873). The obtained sequence was identical to the Italian one (Figure 9) obtained by Carella et al. [10].

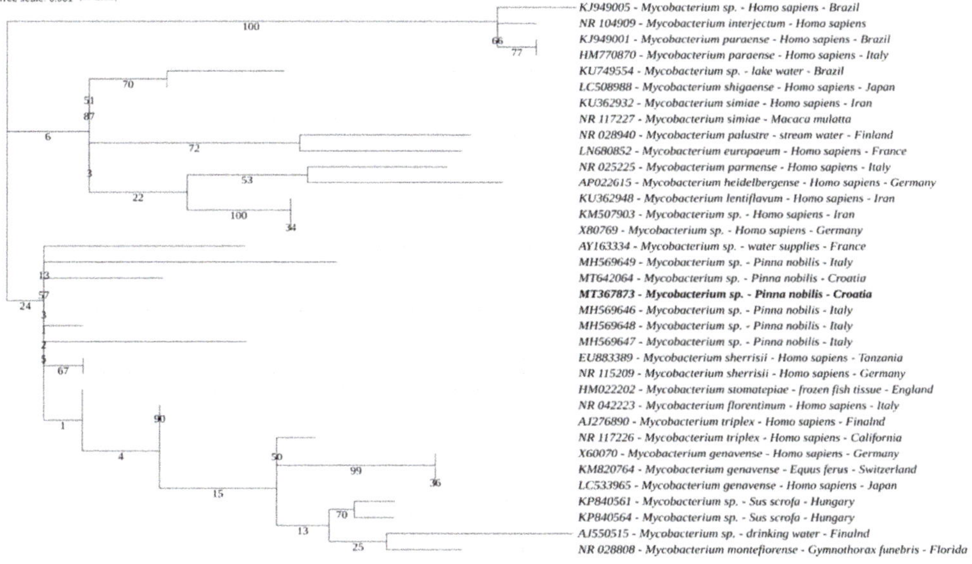

Figure 9. Maximum likelihood phylogenetic tree for *Mycobacterium* spp. The numbers on the branches represent bootstrap values.

4. Discussion

The results of this study are in line with the results obtained by other Croatian researchers [14,15] and clearly show that MMEs have spread from the south to the middle parts of the Eastern Adriatic regions, as well as, ultimately, to the north. We first detected *Haplosporidium pinnae* in association with mortalities on the Croatian Adriatic coast in individuals collected in mid-April 2019 on the outer side of the Mljet Island, situated on the most southern point. At the same time, there were no mortality events among populations in the inner parts of the Mljet Island, and all the sampled animals tested negative for *H. pinnae*, but two individuals from the Small Lake tested positive for *Mycobacterium* spp., as shown in Table 1. Then, in May, mortalities and detection of samples from Elafiti Island (which is a bit more to the north) that were positive for *H. pinnae* and *Mycobacterium* spp. were reported [14]. During the summer period, mortalities were reported at different sites along the coast and close to the islands, mostly in the southern and middle parts of the Eastern Adriatic coast [15]. Interestingly, the data from the mentioned reports, as well as notifications from fishermen, divers, and other parties, indicated that the mortalities primarily affected habitats on the outer parts of the islands and were spreading towards sites that were closer to the mainland—from southern sites to northern sites. In 2019, there were no reports of mortalities in the northern part of the Eastern Adriatic coast, around the Istrian peninsula. The results of the laboratory analysis of the sampled animals supported the absence of both pathogens in the northern regions. Unfortunately, less than half a year later, during May 2020, the pathogens reached the northern sites along the Istrian peninsula and caused high mortalities of the natural populations (Table 1), as indicated by the diagnosis with *H. pinnae* and *Mycobacterium* spp.

Generally, it seems that the mortalities were triggered by an increase in the temperature in the spring, and they reached a high intensity when the sea had a high temperature during the summer. The results of the pilot study undertaken in the Small Lake of the Mljet National Park (Figure 2) support the hypothesis that the increase in mortalities was influenced by high temperatures. The MMEs did not affect populations in the inner parts of the lakes of the National Park from April until the beginning of August, when high mortalities were observed in the Big Lake. This was strange since the distance from the site of the first record of *H. pinnae* at Cape Lenga (Figure 2, sampling point 5) is about 4 km, and the water enters into the lakes from the open sea. At the same time, the MMEs spread to more distant sites in the Middle Adriatic. It seems that the inlet into the lakes does not allow the entry of large quantities of water, which also creates a barrier to the entry of pathogens and causes them to proliferate slowly. Most likely, the high temperature of the sea, which reached its maximum in August, strongly supported the extremely quick spread of the pathogen in the Big Lake and also affected the more isolated population in the Small Lake. The observations from the pilot study indicate that some regularity exists in the spread of the disease. The mortalities of individuals along T1, which was closer to the inlet of the water from the Big Lake, occurred earlier than those along T2. It was also observed that the spread of the disease was not linear, but rather diffuse, affecting primarily weaker individuals and spreading to others in close proximity in the dense population. A higher mortality rate was also observed in the shallower parts of the studied area. The difference related to the depth might be attributed to the higher water temperature in the shallower areas and the depletion of oxygen, which could additionally worsen the environmental conditions for the host. Additionally, it seems that higher water temperatures favor the propagation of both of the detected pathogens. *Mycobacterium* spp. require a higher water temperature for growth, as it is known that a representative of this genus, *Mycobacterium marinum*, a pathogen of marine fish, usually causes disease during warmer parts of the year with water temperatures above 25 °C [41]. Moreover, *Mycobacterium* spp. are opportunistic bacteria that cause chronic diseases, and it should be possible to detect their presence in healthy individuals long before they cause granulomatous lesions, as observed by Carella et al. [10] and Latos et al. [12]. Unfortunately, we analyzed very few samples using histology, and in our scarce histological slides, we did not observe the lesions described by the aforementioned authors [10,12]. Interestingly, in the samples of mantle biopsies collected during the pilot study at the end of August, we detected *H. pinnae* in four out of the 20 tested animals and *Mycobacterium* spp. in two out the 20 animals, which could suggest that the parasite was more representative and more responsible for the MMEs compared to *Mycobacterium* spp. It should be emphasized that the pen shells in the Small Lake of the Mljet National Park did not show increased mortality for 6 months, although it was confirmed that they were infected with *Mycobacterium* spp. in April. Infection with *Mycobacterium* spp. does not appear to be sufficient to trigger a mass mortality event among pen shells, but its co-occurrence with *H. pinnae* contributes to high mortality. Similarly, Box et al. [42] concluded that high mortalities occur in noble pen shells that have been infected with *H. pinnae* and additionally aggravated by co-infection with *Mycobacterium* spp. or other Gram-negative bacteria.

From the results of morphometry, it was obvious that all the marked animals were adults, with the smallest reaching a maximum height of 37.9 cm. It seems that the older individuals were more susceptible to infection with *H. pinnae*, as previously reported by Vázquez-Luis et al. [7], as well as to cases of infection with other genera of *Haplosporidium*, where mortalities occurred in older individuals concurrently with the highest levels of infection [43].

The peak prevalence of the mortalities of Eastern oyster (*Crassostrea virginica*) caused by *Haplosporidium nelsoni* in Delaware Bay, USA [43] was reported in autumn. The mortalities observed in the pilot study with the pen shells peaked during the autumn and slowly decreased during the winter, similarly to those caused by *H. nelsoni*. From the Kaplan–Meier survival estimates (Figure 6), it was visible that the mortalities increased gradually and that the most numerous mortalities occurred at the end of September and again at the end of October. Unfortunately, only four animals survived until the end of November. Undoubtedly, the pathogens identified in our research were those detected as the cause of mortalities throughout the Mediterranean region. The phylogenetic analysis of our isolates of *H. pinnae* showed 99.8% similarity (Figure 8) with those described in the Ionian Sea [9] and 99.7% similarity with Spanish isolates [8]. The phylogenetic analysis showed that the 16S sequences of mycobacteria detected in *P. nobilis* were more distant from each other than some sequences of mycobacteria belonging to different species. For example, *M. sherrisii*, *M. stomatepiae*, *M. florentinum*, *M. triplex*, and *Mycobaterium* spp. from *P. nobilis* from Croatia (MT367873) and Italy (MH569646) are more similar to each other than the two strains of *Mycobaterium* spp. from *P. nobilis* from Italy (MH569647 and MH569649) (Figure 9). This suggests that the detected mycobacteria from *P. nobilis* belong to a heterogeneous group and are not members of the same species.

There are still doubts about what enabled the quick spread of the pathogen through the Mediterranean basin. It was undoubtable that the pathogen entered into the Adriatic Sea and spread from the south to the north by following the main flows of the currents [44]. It is hard to believe that only water currents carry the pathogen, and the pathogen could probably spread through pelagic larval stages of other intermediate hosts. Nevertheless, the histological finding of all the stages of *H. pinnae* in the same animal—uninucleate cells observed in the mantle, gills, gonads, digestive gland, and connective tissue, followed by plasmodia in the connective tissue of the digestive gland, and sporocysts in the epithelium of the digestive tubules (Figure 7a–f)—contributed to the direct transmission from one to another individual, as postulated by Catanese et al. [8]. The quick spread of mortalities within the dense population in the pilot study in the Mljet National Park suggests the direct transmission of *H. pinnae*. Since it is known that spores of *Haplosporidium* are persistent in the environment [45], the involvement of an intermediate host that is abundant in the marine environment and is capable of quick movement through water should be considered, and future studies should aim to prove this hypothesis.

Further joint efforts of marine biologists and invertebrate health experts should be engaged in aiming to understand why only adult specimens were present in particular marine areas, and the reasons were for the increased susceptibility of the noble pen shells to pathogens.

Supplementary Materials: The following are available online at https://www.mdpi.com/article/10.3390/jmse9070764/s1. Table S1: Morphometric characteristics of noble pen shells along the transect 1 (T1); Table S2: Morphometric characteristics of noble pen shells along the transect 2 (T2).

Author Contributions: Conceptualization, Ž.M.; data curation, M.B.; formal analysis, Ž.P. and I.G.Z.; funding acquisition, B.H.; investigation, A.P., O.P., and I.S.; methodology, D.O.; writing—original draft, S.Z. All authors have read and agreed to the published version of the manuscript.

Funding: This research received no external funding.

Institutional Review Board Statement: All the sampling was carried out with the permission of the Ministry of Environmental Protection and Energy (CLASS: UP/1-612-17/18-48/172, No. 517-05-1-1-18-4 issued on 21 December 2018 and CLASS: UP/1-612-07/19-48/193, No. 517-05-1-1-19-3 issued on 11 September 2019).

Data Availability Statement: All the data are available upon reasonable request from the corresponding author.

Acknowledgments: The authors kindly acknowledge Angela Bradarić for the technical support in the field and the laboratory work, Pavel Ankon for the instructions on non-lethal sampling, and Ana Car and Iris Duplić Radić for sharing the salinity data in the Small Lake during the pilot study. Special thanks go to Isabelle Arzul and the EURL for mollusk diseases (IFREMER, La Tremblade) for their confirmation of the detection of *Haplosporidium pinnae*.

Conflicts of Interest: The authors declare no conflict of interest.

References

1. Zavodnik, D.; Hrs-Brenko, M.; Legas, M. Synopsis on the fan shell *Pinna nobilis* L. in the eastern Adriatic Sea. Les Espèces Marines à protegér en Méditerranée. Gis Posidonie. *Publ. Marseille* **1991**, 169–178.
2. Cabanellas-Reboredo, M.; Vazquez-Luis, M.; Mourre, B.; Álvarez, E.; Deudero, S.; Amores, Á.; Addis, P.; Bal-lesteros, E.; Barrajon, A.; Coppa, S.; et al. Tracking a mass mortality outbreak of pen shell *Pinna nobilis* populations: A collaborative effort of scientists and citizens. *Sci. Rep.* **2019**, *9*, 1–11. [CrossRef]
3. Tsatiris, A.; Papadopoulos, V.; Makri, D.; Topouzelis, K.; Manoutsoglou, E.; Hasiotis, T.; Katsanevakis, S. Spatial distribution, abundance and habitat use of the endemic Mediterranean fan mussel *Pinna nobilis* in Gera Gulf, Lesvos (Greece): Comparison of design-based and model-based approaches. *Mediterr. Mar. Sci.* **2018**, *19*, 642–655. [CrossRef]
4. Katsanevakis, S. Growth and mortality rates of the fan mussel *Pinna nobilis* in Lake Vouliagmeni (Korin-thiakos Gulf, Greece): A generalized additive modelling approach. *Mar. Biol.* **2007**, *152*, 1319–1331. [CrossRef]
5. Richardson, C.A.; Peharda, M.; Kennedy, H.; Kennedy, P.; Onofri, V. Age, growth rate and season of recruitment of *Pinna nobilis* (L) in the Croatian Adriatic determined from Mg:Ca and Sr:Ca shell profiles. *J. Exp. Mar. Biol. Ecol.* **2004**, *299*, 1–16. [CrossRef]
6. Deudero, S.; Grau, A.; Vázquez-Luis, M.; Álvarez, E.; AlOmar, C.; Hendriks, I.E. Reproductive investment of the pen shell *Pinna nobilis* Linnaeus, 1758 in Cabrera National Park (Spain). *Mediterr. Mar. Sci.* **2017**, *18*, 271. [CrossRef]
7. Vázquez-Luis, M.; Álvarez, E.; Barrajón, A.; García-March, J.R.; Grau, A.; Hendriks, I.E.; Jiménez, S.; Kersting, D.; Moreno, D.; Pérez, M.; et al. S.O.S. *Pinna nobilis*: A Mass Mortality Event in Western Mediterranean Sea. *Front. Mar. Sci.* **2017**, *4*, 220. [CrossRef]
8. Catanese, G.; Grau, A.; Valencia, J.M.; March, J.R.G.; Vázquez-Luis, M.; Alvarez, E.; Deudero, S.; Darriba, S.; Carballal, M.J.; Villalba, A. *Haplosporidium pinnae* sp. Nov., a haplosporidan parasite associated with mass mortalities of the fan mussel, *Pinna nobilis*, in the Western Mediterranean Sea. *J. Invertebr. Pathol.* **2018**, *157*, 9–24. [CrossRef]
9. Panarese, R.; Tedesco, P.; Chimienti, G.; Latrofa, M.S.; Quaglio, F.; Passantino, G.; Buonavoglia, C.; Gustinelli, A.; Tursi, A.; Otranto, D. *Haplosporidium pinnae* associated with mass mortality in endangered *Pinna nobilis* (Linnaeus 1758) fan mussels. *J. Invertebr. Pathol.* **2019**, *164*, 32–37. [CrossRef]
10. Carella, F.; Aceto, S.; Pollaro, F.; Miccio, A.; Iaria, C.; Carrasco, N.; Prado, P.; De Vico, G. A mycobacterial disease is associated with the silent mass mortality of the pen shell *Pinna nobilis* along the Tyrrhenian coast-line of Italy. *Sci. Rep.* **2019**, *9*, 2725. [CrossRef]
11. Katsanevakis, S. The cryptogenic parasite *Haplosporidium pinnae* invades the Aegean Sea and causes the collapse of *Pinna nobilis* populations. *Aquat. Invasions* **2019**, *14*, 150–164. [CrossRef]
12. Lattos, A.; Giantsis, I.A.; Karagiannis, D.; Michaelidis, B. First detection of the invasive *Haplosporidian* and *Mycobacteria parasites* hosting the endangered bivalve *Pinna nobilis* in Thermaikos Gulf, North Greece. *Mar. Environ. Res.* **2020**, *155*, 104889. [CrossRef] [PubMed]

13. Tiscar, P.G.; Rubino, F.; Fanelli, G.; Paoletti, B.; Della Salda, L. Mass Mortality of the Fan Mussel *Pinna nobilis* in Apulia (Ionian Sea) Caused by *Haplosporidium pinnae*. In Proceedings of the 42nd CIESM Congress, Cascais, Portugal, 7–11 October 2019.
14. Čižmek, H.; Čolić, B.; Gračan, R.; Grau, A.; Catanese, G. An emergency situation for pen shells in the Mediterranean: The Adriatic Sea, one of the last *Pinna nobilis* shelters, is now affected by a mass mortality event. *J. Invertebr. Pathol.* **2020**, *173*, 107388. [CrossRef] [PubMed]
15. Šarić, T.; Župan, I.; Aceto, S.; Villari, G.; Palić, D.; De Vico, G.; Carella, F. Epidemiology of Noble Pen Shell (*Pinna nobilis* L. 1758) Mass Mortality Events in Adriatic Sea Is Characterised with Rapid Spreading and Acute Disease Progression. *Pathogens* **2020**, *9*, 776. [CrossRef] [PubMed]
16. Künili, I.E.; Gürkan, S.E.; Aksu, A.; Turgay, E.; Çakir, F.; Gürkan, M.; Altinağaç, U. Mass mortality in endangered fan mussels *Pinna nobilis* (Linnaeus 1758) caused by co-infection of *Haplosporidium pinnae* and multiple *Vibrio* infection in Çanakkale Strait, Turkey. *Biomarkers* **2021**, *26*, 450–461. [CrossRef]
17. International Union for Conservation of Nature (IUCN). Mediterranean Noble Pen Shell Crisis (Pinna nobilis)-January 2020 Update. Available online: https://www.iucn.org/news/mediterranean/202001/mediterranean-noble-pen-shell-crisis-pinna-nobilis-january-2020-update (accessed on 10 March 2021).
18. Kersting, D.; Benabdi, M.; Čižmek, H.; Grau, A.; Jimenez, C.; Katsanevakis, S.; Öztürk, B.; Tuncer, S.; Tunesi, L.; Vázquez-Luis, M.; et al. *Pinna Nobilis. The IUCN Red List of Threatened Species 2019*; IUCN Red List: London, UK, 2019. [CrossRef]
19. Pavlinec, Ž.; Zupičić, I.G.; Oraić, D.; Petani, B.; Mustać, B.; Mihaljević, Ž.; Beck, R.; Zrnčić, S. Assessment of predominant bacteria in noble pen shell (*Pinna nobilis*) collected in the Eastern Adriatic Sea. *Environ. Monit. Assess.* **2020**, *192*, 1–10. [CrossRef]
20. Carella, F.; Antuofermo, E.; Farina, S.; Salati, F.; Mandas, D.; Prado, P.; Panarese, R.; Marino, F.; Fiocchi, E.; Pretto, T.; et al. In the wake of the ongoing mass mortality events: Co-occurrence of *Mycobacterium*, *Haplosporidium* and other pathogens in *Pinna nobilis* collected in Italy and Spain (Mediterranean Sea). *Front. Mar. Sci.* **2020**, *7*, 48. [CrossRef]
21. Prado, P.; Carrasco, N.; Catanese, G.; Grau, A.; Cabanes, P. Presence of *Vibrio mediterranei* associated to major mortality in stabled individuals of *Pinna nobilis* L. *Aquaculture* **2020**, *519*, 734899. [CrossRef]
22. Lattos, A.; Bitchava, K.; Giantsis, I.; Theodorou, J.; Batargias, C.; Michaelidis, B. The Implication of *Vibrio* Bacteria in the Winter Mortalities of the Critically Endangered *Pinna nobilis*. *Microorganisms* **2021**, *9*, 922. [CrossRef]
23. Scarpa, F.; Sanna, D.; Azzena, I.; Mugetti, D.; Cerruti, F.; Hosseini, S.; Cossu, P.; Pinna, S.; Grech, D.; Cabana, D.; et al. Multiple Non-Species-Specific Pathogens Possibly Triggered the Mass Mortality in *Pinna nobilis*. *Life* **2020**, *10*, 238. [CrossRef]
24. Lattos, A.; Giantsis, I.A.; Karagiannis, D.; Theodorou, J.A.; Michaelidis, B. Gut Symbiotic Microbial Communities in the IUCN Critically Endangered *Pinna nobilis* Suffering from Mass Mortalities, Revealed by 16S rRNA Amplicon NGS. *Pathogens* **2020**, *9*, 1002. [CrossRef]
25. Orepić, N.; Vidmar, J.; Zahtila, E.; Zavodnik, D. A marine benthos survey in the lakes of the National park Mljet (Adriatic Sea). *Period. Biol.* **1997**, *99*, 229–245.
26. Šiletić, T.; Peharda, M. Population study of the fan shell *Pinna nobilis* L. in Malo and Veliko Jezero of the Mljet National Park (Adriatic Sea). *Sci. Mar.* **2003**, *67*, 91–98. [CrossRef]
27. Peharda, M.; Hrs-Brenko, M.; Bogner, D.; Onofri, V.; Benović, A. Spatial distribution of live and dead bivalves in saltwater lake Malo Jezero (Mljet National Park). *Period. Biol.* **2002**, *104*, 115–122.
28. Howard, D.; Lewis, E.; Keller, J.; Smith, C. *Histological Techniques for Marine Bivalve Mollusks and Crustaceans*, 2nd ed.; NOAA (National Ocean Service): Oxford, MD, USA, 2004; 60p.
29. Garcia-March, J.R.; Vicente, N. Protocol to Study and Monitor Pinna nobilis Populations within Marine Protected Areas. Report of the Malta Environment and Planning Authority (MEPA) and MedPAN-Interreg IIIC-Project. 2006. Available online: https://drive.google.com/file/d/131bRQrcQQ62NUep1V1J3f_W33JAmANbO/view (accessed on 5 September 2019).
30. Sanna, D.; Cossu, P.; Dedola, G.L.; Scarpa, F.; Maltagliati, F.; Castelli, A.; Franzoi, P.; Lai, T.; Cristo, B.; Curini-Galletti, M.; et al. Mitochondrial DNA Reveals Genetic Structuring of *Pinna nobilis* across the Mediterranean Sea. *PLoS ONE* **2013**, *8*, e67372. [CrossRef] [PubMed]
31. Dudley, W.N.; Wickham, R.; Coombs, M.N. An Introduction to Survival Statistics: Kaplan-Meier Analysis. *J. Adv. Pr. Oncol.* **2016**, *7*, 91–100. [CrossRef]
32. Morton, B.; Puljas, S. An improbable opportunistic predator: The functional morphology of *Pinna nobilis* (Bivalvia: Pterioida: Pinnidae). *J. Mar. Biol. Assoc. UK* **2019**, *99*, 359–373. [CrossRef]
33. Renault, T.; Stokes, N.; Chollet, B.; Cochennec, N.; Berthe, F.; Gerard, A.; Burreson, E. Haplosporidiosis in the Pacific oyster *Crassostrea gigas* from the French Atlantic coast. *Dis. Aquat. Org.* **2000**, *42*, 207–214. [CrossRef]
34. Telenti, A.; Marchesi, F.; Balz, M.; Bally, F.; Böttger, E.C.; Bodmer, T. Rapid identification of mycobacteria to the species level by polymerase chain reaction and restriction enzyme analysis. *J. Clin. Microbiol.* **1993**, *31*, 175–178. [CrossRef] [PubMed]
35. Böddinghaus, B.; Rogall, T.; Flohr, T.; Blöcker, H.; Böttger, E.C. Detection and identification of mycobacteria by amplification of rRNA. *J. Clin. Microbiol.* **1990**, *28*, 1751–1759. [CrossRef] [PubMed]
36. Katoh, K.; Standley, D.M. MAFFT Multiple Sequence Alignment Software Version 7: Improvements in Performance and Usability. *Mol. Biol. Evol.* **2013**, *30*, 772–780. [CrossRef]
37. Benson, D.A.; Cavanaugh, M.; Clark, K.; Mizrachi, I.K.; Lipman, D.J.; Ostell, J.; Sayers, E.W. GenBank. *Nucleic Acids Res.* **2012**, *41*, D36–D42. [CrossRef] [PubMed]

38. Altschul, S.F.; Gish, W.; Miller, W.; Myers, E.W.; Lipman, D.J. Basic local alignment search tool. *J. Mol. Biol.* **1990**, *215*, 403–410. [CrossRef]
39. Kumar, S.; Stecher, G.; Li, M.; Knyaz, C.; Tamura, K. MEGA X: Molecular evolutionary genetics analysis across computing platforms. *Mol. Biol. Evol.* **2018**, *35*, 1547–1549. [CrossRef] [PubMed]
40. Letunic, I.; Bork, P. Interactive Tree of Life (iTOL) v4: Recent updates and new developments. *Nucleic Acids Res.* **2019**, *47*, W256–W259. [CrossRef] [PubMed]
41. Hashish, E.; Merwad, A.-R.; Elgaml, S.; Amer, A.; Kamal, H.; Elsadek, A.; Marei, A.; Sitohy, M. *Mycobacterium marinum* infection in fish and man: Epidemiology, pathophysiology and management; a review. *VetIQ Petcare* **2018**, *38*, 35–46. [CrossRef] [PubMed]
42. Box, A.; Capó, X.; Tejada, S.; Catanese, G.; Grau, A.; Deudero, S.; Sureda, A.; Valencia, J.M. Reduced Antioxidant Response of the Fan Mussel *Pinna nobilis* Related to the Presence of *Haplosporidium pinnae*. *Pathogens* **2020**, *9*, 932. [CrossRef]
43. Arzul, I.; Carnegie, R.B. New perspective on the haplosporidian parasites of molluscs. *J. Invertebr. Pathol.* **2015**, *131*, 32–42. [CrossRef]
44. Mihanović, H.; Cosoli, S.; Vilibić, I.; Ivanković, D.; Dadić, V.; Gačić, M. Surface current patterns in the northern Adriatic extracted from high-frequency radar data using self-organizing map analysis. *J. Geophys. Res. Space Phys.* **2011**, *116*, 08033. [CrossRef]
45. Haskin, H.H.; Andrews, J.D. Uncertainties and Speculations about the Life Cycle of the Eastern Oyster Pathogen *Haplosporidium nelsoni* (MSX). In *Disease Processes in Marine Bivalve Molluscs*; American Fisheries Society: Bethesda, MD, USA, 1988; Available online: https://scholarworks.wm.edu/cgi/viewcontent.cgi?article=1117&context=vimsbooks (accessed on 5 March 2021).

Journal of
*Marine Science
and Engineering*

Bonamia exitiosa in European Flat Oyster (*Ostrea edulis*) on the Croatian Adriatic Coast from 2016 to 2020

Dražen Oraić [1,*], Relja Beck [1], Željko Pavlinec [1], Ivana Giovanna Zupičić [1], Ljupka Maltar [2], Tihana Miškić [2], Žaklin Acinger-Rogić [2] and Snježana Zrnčić [1]

[1] Croatian Veterinary Institute, 10000 Zagreb, Croatia; beck@veinst.hr (R.B.); pavlinec@veinst.hr (Ž.P.); zupicic@veinst.hr (I.G.Z.); zrncic@veinst.hr (S.Z.)
[2] Croatian Ministry of Agriculture-Veterinary and Food Safety Directorate General, 10000 Zagreb, Croatia; ljupka.maltar@mps.hr (L.M.); tihana.miskic@mps.hr (T.M.); zaklin.acinger@mps.hr (Ž.A.-R.)
* Correspondence: oraic@veinst.hr

Abstract: The annual production of European flat oysters (*Ostrea edulis*) in Croatia is about 50 to 65 tons, and it has a long tradition. All Croatian oyster farms are subjected to the national surveillance program aiming to detect the presence of *Bonamia ostreae* and *Marteilia refringens* according to the Council Directive 2006/88/EC. Within the surveillance program, the first findings of the parasite *Bonamia* spp. occurred in 2016 in two production areas in the north and south of the Eastern Adriatic coast. The repeated findings of the parasite were noted up to 2020 but also on two additional sites in the north. The parasite was detected by cytological analysis of stained heart smears, histological examination, and PCR. PCR positive samples were sequenced for SSU rDNA gene, and BLAST analysis confirmed infection with *Bonamia exitiosa*. Attempts to prove the Pacific oyster as a putative vector of the parasite failed. The infection prevalence from 2016 until 2020 ranged from 3.3 to 20% in different sites. No mortalities were reported from the infected sites, and it seemed that infection of flat oysters with *B. exitiosa* did not affect their health. The study has not shown the source and way of infection spread, which imposes the need for more comprehensive molecular and epidemiological studies.

Keywords: *Bonamia exitiosa*; bonamiosis; Croatian Adriatic coast; *Ostrea edulis*; prevalence; surveillance

Citation: Oraić, D.; Beck, R.; Pavlinec, Ž.; Zupičić, I.G.; Maltar, L.; Miškić, T.; Acinger-Rogić, Ž.; Zrnčić, S. *Bonamia exitiosa* in European Flat Oyster (*Ostrea edulis*) on the Croatian Adriatic Coast from 2016 to 2020. *J. Mar. Sci. Eng.* **2021**, *9*, 929. https://doi.org/10.3390/jmse9090929

Academic Editor: Patrizia Pagliara

Received: 23 June 2021
Accepted: 21 August 2021
Published: 27 August 2021

Publisher's Note: MDPI stays neutral with regard to jurisdictional claims in published maps and institutional affiliations.

1. Introduction

The production of European flat oyster (*Ostrea edulis*) has a long tradition and economic significance in many countries of the Mediterranean basin and the East Atlantic coast. Along the Croatian coast, flat oysters have been cultivated for centuries, and in recent decades, the annual production has ranged from 50 to 65 t [1]. Production of *O. edulis* has stagnated since the 70s of the last century as a result of the introduction of pathogens into the susceptible population. Therefore, diseases have caused large losses and dramatically affected production [2–4]. One of these pathogens is the protozoan parasite *Bonamia ostreae*, belonging to the clade "microcell" within Haplosporidia [5]. Various methods have been developed to identify and confirm the presence of parasites of the genus *Bonamia*, but each of them has certain limitations. Microscopy of gill and heart-stained tissue impressions and histological slides are methods to identify and confirm *Bonamia* species. Identifying parasite species due to morphological similarity is time-consuming and requires experience and expertise. However, both these methods have low sensitivity and specificity [6–9], similar to TEM, which seems to be insufficient in the identification of species [10]. Diagnostic sensitivity has significantly improved with the implementation of molecular methods, such as PCR [11,12] or in situ hybridization.

Based on molecular studies of small ribosomal subunit rRNA genes, the genera *Bonamia* is phylogenetically positioned into a clade microcell within the genus *Haplosporidia*, and within it, includes four species: *B. ostreae*, *B. exitiosa*, *B. roughleyi*, and *B. perspora* [13–16].

All species of *Bonamia* cause infection of flat oyster hemocytes, leading to death [12]. In some cases, parasites are present in low prevalence and with little impact on the flat oyster population [17]. It has been reported that flat oysters exposed to parasites over a long period develop a certain degree of resistance [18]. *Bonamia exitiosa* was for the first time reported as the cause of the Chilean oyster's (*Ostrea chilensis*) high mortality in New Zealand when only 9% of affected flat oysters survived [19]. Catastrophic and rapid mortality was caused by the same parasite on triploids of the Suminoe oyster (*Crassostrea ariakensis*), introduced for experimental cultivation in 2003 in Bogue Sound, North Carolina [20], in an area where the crested oyster (*Ostreola equestris*) was an indigenous species [15].

This paper presents the findings, distribution, and prevalence of the *Bonamia exitiosa* in the samples of flat oyster analyzed within the national surveillance program for the infection with *Bonamia* parasites.

2. Materials and Methods

2.1. Sampling

The sampling points within production areas were determined through an intensive monitoring program in the period from 1998 to 2000. Each year, according to the national surveillance program, authorized veterinarians sampled 30 adult flat oysters on each of nine sampling points along the Croatian Adriatic coast (Figure 1, Table 1). The samples were packed in a transport refrigerator and delivered to the laboratory within 24–36 h.

Figure 1. Map showing European flat oyster (*O. edulis*) production areas with sampling points included in the national surveillance plan. A production area in the south with five sampling points and three production areas in the north Croatian Adriatic Sea tested positive (red numbers) for *B. exitiosa* in 2016–2020. The numbers refer to production areas/sampling sites listed in Table 2.

Table 1. Positive findings of *Bonamia exitiosa* within sampling area showing sampling dates, sampling points, and prevalence using different diagnostic techniques in Croatia during the period 2016–2020.

Species	Flat Oyster (*Ostrea edulis*)											Pacific Oyster (*Magallana gigas*)			
Production Area	Mali Ston Bay		Medulin Bay		Lim Bay		Savudrija Bay		Marina Bay			Medulin Bay		Lim Bay	
	Date M/Y	N^o	Date M/Y	N^o	Date M/Y	N^o	Date M/Y	N^o	Date M/Y	N^o		Date M/Y	N^o	Date M/Y	N^o
Date and number of sample animals	05/2016	150 *	05/2016	30 *	05/2016	30	05/2016	30	11/2016	30					
	11/2016	150	11/2016	30 *	09/2016	30	10/2016	30	06/2017	30					
	05/2017	150	07/2017	30 *	05/2017	30	05/2017	30	05/2018	30		07/2017	18		
	11/2017	150	11/2017	30 *	06/2018	30 *	06/2018	30 *	09/2018	30					
	11/2018	150	10/2018	30	09/2019	30	10/2019	30	08/2019	30				09/2018	30
	10/2020	150	10/2020	30 *	09/2020	30	09/2020	30 *	08/2020	30					
Totally		900		180		180		180		180			18		30

* Samples with positive finding of *B. exitiosa*.

Table 2. Total number of oysters analyzed for the presence of *Bonamia* spp. in the period 2016–2020.

Date of Sampling (M/Y)	Production Area/Name of Sampling Site	Number of Sampled Animals	Sampled Species	Results of Laboratory Analysis					
				Heart Imprints		PCR		Real Time PCR	
	MALOSTONSKI BAY AREA			+-ve	%	+-ve	%	+-ve	%
05/2016	6 * Mali Ston	30	*Ostrea edulis*	3	10	5	16.7	n/d	-
05/2016	2 * Brijesta	30	*Ostrea edulis*	3	10	6	2.,0	n/d	-
05/2016	5 * Bistrina	30	*Ostrea edulis*	2	6.7	2	6.7	n/d	-
05/2016	3 * Bjejevica	30	*Ostrea edulis*	2	6.7	3	10.0	n/d	-
05/2016	4 * Sutvid	30	*Ostrea edulis*	1	3.3	1	3.3	n/d	-
	MEDULINSKI BAY								
05/2016	7 * Medulinski Bay	30	*Ostrea edulis*	2	6.7	3	6.7	n/d	-
11/2016	7 * Medulinski Bay	30	*Ostrea edulis*	1	3.3	1	3.3	n/d	-
07/2017	7 * Medulinski Bay	30	*Ostrea edulis*	2	6.7	3	10.0	n/d	-
11/2017	7 * Medulinski Bay	30	*Ostrea edulis*	1	3.3	1	3.3	n/d	-
11/2017	7 * Medulinski Bay	18	*Magallana gigas*	0	0	0	0	n/d	-
10/2020	7 * Medulinski Bay	30	*Ostrea edulis*	4	13.3	5	16.7	5	16.7
	LIM BAY								
07/2018	1 * Lim Bay	30	*Ostrea edulis*	1	3.3	2	6.7	n/d	-
2018	1 * Lim Bay	30	*Magallana gigas*	0	0	0	0	n/d	-
	SAVUDRIJA BAY								
06/2018	12 * Savudrija Bay	30	*Ostrea edulis*	2	6.7	3	10.0	n/d	-
09/2020	12 * Savudrija Bay	30	*Ostrea edulis*	1	6.7	4	13.3	4	13.3

* labels the number of the sampling site corresponding to the map of sampling sites.

All samples submitted within the surveillance program in the period 2016 to 2020, from the *B. exitiosa* positive sites are listed in Table 2 and sites are shown in Figure 1. Namely, from the site of the Medulin Bay during the mentioned period (2016–2020), 180 flat oysters were submitted; from the Mali Ston Bay with five sampling sites, there were 900 individuals, while 180 individuals were submitted from the sites from Lim Bay, Savudrija Bay, and Marina Bay, respectively.

After the first finding of *B. exitiosa* positive flat oysters in the northern part of the Eastern Adriatic coast, epidemiological surveillance was carried out. It was found that there were some populations of Pacific oyster (*Magallana* = *Crassostrea gigas*) present in the Medulin Bay production area, and during 2017, a sample of 18 animals was collected to examine them as a possible source of infection. In 2018, after the detection of *B. exitiosa* in Lim Bay, another batch of 30 Pacific oyster was tested for the presence of this parasite.

2.2. Tissue Processing, Cytological and Histological Examination

After dissection of each oyster submitted for diagnostics, including Pacific oysters, tissues were collected for cytology, histology, and molecular analysis.

For cytological diagnosis of *Bonamia* spp., a piece of the heart of each oyster was taken and briefly dried on absorbent paper. A series of heart imprints were made in two rows on a glass slide. Imprints of five specimens were made on one slide and stained with Hemacolor® staining kit according to manufacturer instructions (Merck, Darmstadt, Germany).

For histological examination, transversal sections of soft tissue, including gills and the digestive gland of each oyster, were placed in histo-cassettes and immersed in Davidson's formalin, alcohol, acetic acid (AFA) fixative. Davidson-fixed tissues were dehydrated through a graded series of ethanols, succeeded by xylene, and embedded in paraffin, sectioned at 3 μm, and mounted on Microme EC 350-2 slides (Thermo Scientific, Waltham, MA, USA). Mounted slides were heated to 60 °C, deparaffinized, and rehydrated in xylene, a graded series of alcohol, and finally, water, followed by staining with hematoxylin and eosin (H&E).

For detection of *Bonamia* spp. by PCR, gill samples were taken and preserved in ethanol 96% until DNA extraction.

Stained tissue impressions and histological sections were examined on a Zeiss Axiskop-2 binocular microscope (Carl Zeiss, Jena, Germany) at 400× and 1000× magnifications with immersion oil.

2.3. Molecular Detection and Characterization

2.3.1. DNA Extraction

DNA used in all molecular analyses was extracted from gill tissues from each animal. From the samples collected from 2016–2019, DNA was extracted using MagMAX CORE Nucleic Acid Purification Kit (Thermo Fisher Scientific, USA) on KingFisher Flex System (Thermo Fisher Scientific, Waltham, MA, USA). From the samples collected in 2020, DNA was extracted using innuPREP AniPath DNA/RNA Kit—IPC16 (Analytik Jena, Germany) on InnuPure C16 touch (Analytik Jena, Jena, Germany). Both methods of extraction were done according to the manufacturer's instructions.

2.3.2. PCR and Sequencing

In the period from 2016 to 2018, a modified method developed by Cochennec et al. [11] for the detection of *Bonamia* sp was used. The reaction mix contained 10 μL of GoTaq G2 Hot Start Master Mix (Promega, Madison, WI, USA), 60 ng of DNA as measured on a DS-11 Series Spectrophotometer (DeNovix, Wilmington, NC, USA), 0.5 μM of each primer (BO 5′ 3′, BOAS 5′ 3′), and nuclease-free water to the final volume of 20 μL. The amplification was performed using ProFlex PCR System (Applied Biosystems, Waltham, MA, USA) with an initial denaturation at 95 °C for 2 min, followed by 35 cycles of denaturation at 94 °C for 1 min, primer annealing at 55 °C for 1 min, elongation at 72 °C for 1 min, and ending with the final elongation step at 72 °C for 7 min. The presence of PCR products was examined by electrophoresis on a QIAxcel system (Qiagen, Hilden, Germany) using the QIAxcel DNA Screening Kit. For species identification, direct Sanger sequencing of the PCR products was performed by Macrogen Europe. Sequence identity was confirmed by BLAST [21].

2.3.3. Real-Time PCR

For the samples from 2019 and 2020, for the simultaneous detection of *B. ostreae* and *B. exitiosa* DNA, a probe-based real-time qPCR assay developed by the European Union Reference Laboratory (EURL) for Mollusc Diseases (Ifremer, La Tremblade, France) was performed [22]. The reaction mix contained 10 µL of GoTaq® Probe qPCR Master Mix, 2X (Promega, Madison, WI, USA), 0.3 µM of each primer (BO2_F 5′ AAATGGCCTCTTCC-CAATCT 3′, BO2_R 5′ CCGATCAAACTAGGCTGGAA 3′, BEa_F 5′ GACTTTGACCATCG-GAAACG 3′, BEa_R 5′ ATCGAGTCGTACGCGAGTCT 3′), 0.2 µM of each double-labelled probe (BO2_probe 5′ HEX—TGACGATCGGGAATGAACGC—BHQ-1 3′, BEa_probe 5′ FAM—GGCAGCGAATCGATGGGAAT—BHQ-1 3′), 25 ng of template DNA as measured on a DS-11 Series Spectrophotometer (DeNovix, Wilmington, USA), and nuclease-free water to the final volume of 20 µL. The amplification was performed on qTower^3 thermal cycler (Analytik Jena, Jena, Germany) with the following program: 95 °C for 2 min for polymerase activation, followed by 40 cycles of denaturation at 95 °C for 15 s and annealing at 60 °C for 30 s. The fluorescence was recorded at the end of each cycle with HEX and FAM filters. To detect possible differences, all positive samples were amplified and sequenced using the same method described above.

2.3.4. Sequencing and Phylogeny

Obtained sequences were analyzed using Geneious Prime 2020.0.1 software. For the phylogenetic analysis of the obtained sequences, similar sequences available in Gen-Bank [23] were used and identified using BLAST [21]. Sequences were aligned with MAFFT v7.388 [24] using default parameters. Phylogenetic analysis was performed in MEGA 10.0.5 [25]. The best model was selected based on the lowest BIC score. The maximum likelihood tree was calculated with the following settings: K2+G substitution model, five discrete gamma categories, used all sites, and SPR level 5 heuristic method with very strong branch swap filter. Phylogeny was tested by the bootstrap method with 1000 replications. The obtained tree was visualized in iTOL [26]. Four sequences belonging to *Minchinia* spp. were used as an outgroup.

3. Results

Until 2016, the results of cytological examination of heart imprints, histological slides of oysters' organs, and PCR tested negative for the presence of parasite *Bonamia* spp. There were no reported mortalities, and there were no symptoms of the disease on the oysters from the production areas included in the surveillance program. In the May of 2016, for the first time, there were positive findings of the parasite *Bonamia exitiosa* from two production areas (Figure 1, Table 2).

The first one was in the north of the Eastern Adriatic Coast in Medulin Bay where 3 out of 30 animals tested positive for *Bonamia* sp. using conventional PCR and heart imprints, and the second one was in the south of the Croatian Adriatic Sea in the Mali Ston Bay. In all five sampling points, there were *Bonamia* sp. positive samples; (in Mali Ston and Brijesta, 3 out of 30, respectively, by heart imprints and five and six using conventional PCR; in Bistrina and Bjejevica, 2 out of 30, respectively, in imprints and two and three by PCR; and Sutvid, 1 out of 30 by each technique. Following the first finding in Medulin Bay in May, 1 out of 30 animals sampled in November of the same year tested positive. In 2017, positive samples were detected again in Medulin Bay in May and November and also in October of 2020. In 2019, a conventional PCR was replaced with a new molecular method, real-time PCR developed by EURL for mollusc diseases, and validated through interlaboratory proficiency testing as a screening method in the surveillance program. All samples analyzed in 2019 and 2020 were tested using real-time PCR. Therefore, samples collected in Medulin Bay in October 2020 were screened using real-time PCR, and positive samples were tested using conventional PCR and submitted for sequencing. In Mali Ston Bay, there have been no positive findings since 2016. In June 2018, there were positive findings of *B. exitiosa* in 2 out of 30 flat oysters sampled in the Lim Bay detected by imprints and three by PCR and in the same number by each method (two by imprints and three by PCR out of 30 flat oysters) sampled in the Savudrija Bay, two sampling points in the Northeastern Adriatic coast. In Savudrija Bay, in 2020, 4 out of 30 flat oysters were found positive by real-time PCR, confirmed by conventional PCR and sequencing. However, all samples from Marina Bay tested negative for the presence of *Bonamia* sp. during the whole studied period using heart imprints, conventional and real-time PCR.

All pacific oysters collected in the Medulin Bay (n = 18) and from the Limski Bay (n = 30) tested negative for the presence of *Bonamia* sp by both used techniques.

Microscopic examination of heart imprints revealed mononuclear cells (Figure 2a) and, in fewer numbers, binuclear cells of light blue-stained cytoplasm and red-stained nuclei that were located centrally or near the center in mononuclear stages (Figure 2b).

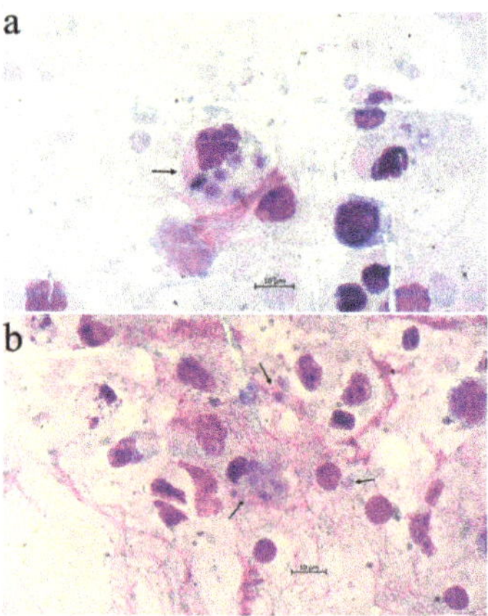

Figure 2. (**a**) Heart imprints of *Ostrea edulis* infected by cells identified as *Bonamia exitiosa* within hemocyte; (**b**) Heart imprints of *Ostrea edulis* infected by mononuclear and binucleated cells identified as *Bonamia exitiosa*. Hemacolor staining, 1000× magnification.

Microscopic examination of histological slides showed weak to moderate infiltration of hemocytes in the connective tissue of gills, digestive glands, and gonads. Weak to medium necrosis with low to medium presence of *Bonamia*-like parasites (Mean size: 1.95 μm) were found in the same tissues. In a small number of infected animals, the connective tissue of the gonads had marked necrotic changes and *Bonamia*-like cells were found (Figure 3a,b).

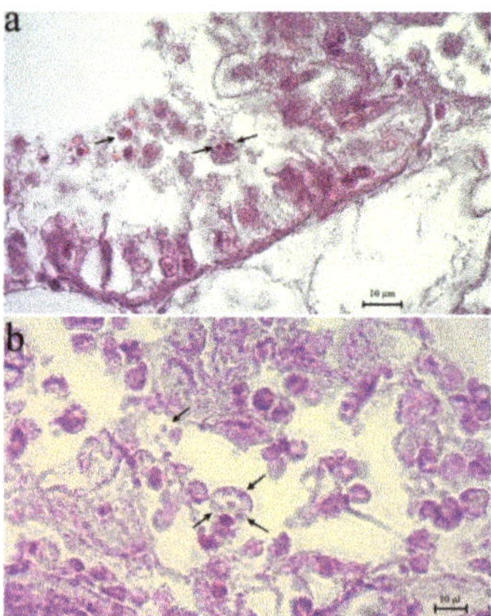

Figure 3. (**a**) Histological section of *Ostrea edulis*; *Bonamia*-like parasite in necrotic tissue of digestive gland (**b**) Gonadal connective tissue cells infected with *Bonamia*-like parasite. H&E staining, 1000× magnification.

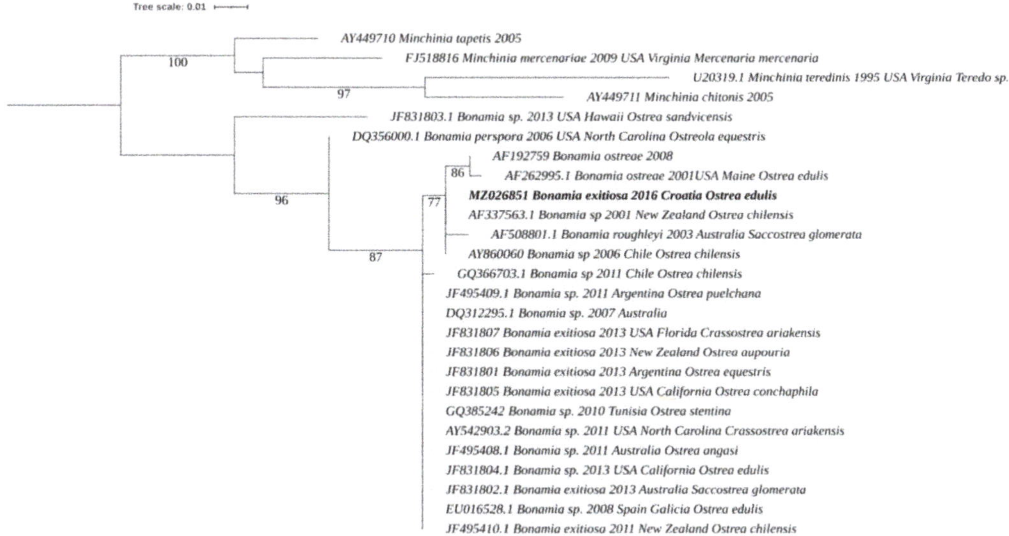

Figure 4. *Bonamia* spp. maximum likelihood phylogenetic tree. Bootstrap values higher than 75 are displayed.

4. Discussion

Until 2010, the diagnostic procedure for detection of *Bonamia* spp. was based on the results of heart imprints and histopathological evaluation of each oyster. Subsequently, since 2010, examinations of stained imprints and PCR screening methods for the same sample have been combined according to a dual detection strategy described by Carnegie and EU legislative [27,28]. In the period from 2010 to the first finding of *Bonamia* spp. positive oysters, 1870 individuals were tested in total.

In the spring of 2016, *Bonamia* parasite was detected for the first time by conventional PCR in the samples collected from the production area in the northern part of the Eastern Adriatic Sea, sampling point 7 (Figure 1). The sample consisted of 30 adult individuals (more than 2-y old), which were reported to be more susceptible to infection with *Bonamia* spp. compared to younger ages [29,30]. The microcellular finding was confirmed in stained heart imprints. The suspicion for the presence of *Bonamia ostreae* was set up, but sequencing revealed the presence of *Bonamia exitiosa*. Shortly following, the samples collected from the production area in the southern part of the Croatian Adriatic coast, the Mali Ston Bay, were submitted for diagnostics. In each of five samples (*n* = 150) from different sampling points, the same diagnostic procedure revealed that 17 flat oysters were PCR positive for *Bonamia* spp. Again, an affiliation of *B. exitiosa* was determined by sequencing of the PCR positive product. It was found that the prevalence of *B. exitiosa* in the different sampling points within Mali Ston Bay notably varied from 3.3% to 20.0% per sampling point (Table 2) when tested by PCR compared to prevalence of 3.3% to 13.3% after evaluation of stained heart imprints. In stained heart imprints, single cells with a central nucleus were observed by microscopic examination. However, outside hemocytes, there was a lower abundance of cells with two nuclei. This corresponds to the previously described imprints findings of *Bonamia exitiosa* in flat oysters [31–34]. Histopathological examination of positive PCR samples of *B. exitiosa* confirmed *Bonamia*-like cells in one-third of samples. In this study, the histological examination was proven as a less sensitive method in line with the finding of Lynch et al., who previously related the low prevalence of parasite to the low sensitivity [35]. It can also mean that environmental conditions were not favorable for the spread and development of infection or that our PCR positive findings were related to the detection of non-infective parasite stages [36].

Interestingly, *B. exitiosa* was detected in all five sampling points in the Mali Ston Bay only in spring of 2016 and never again until 2020, although surveillance was carried out regularly each year, and in total, 750 oysters were examined during this period and tested negative for the presence of *B. exitiosa*. At the same time, in the north of the Adriatic coast, *B. exitiosa* was confirmed for the first time by sequencing and heart imprints in spring 2016, and in following years, it was detected in another two production areas. In Medulin Bay, the findings of *B. exitiosa* continuously occurred from the spring of 2016 to autumn of 2020, except in 2019. During the seasons with positive findings of the parasite, the prevalence of the PCR positive samples varied from 3.3% to 16.7%. Unexpectedly, in 2019, there were no *B. exitiosa* positive/infected oysters, which could be explained with a low number of analyzed samples. There were only 90 individuals submitted for diagnostics of *Bonamia* spp.

It has been experimentally proven that *B. ostreae* favors lower sea temperature for survival [37] and higher prevalence of bonamiosis in oysters, and increased mortality in the colder period with lower sea temperatures has been described. On the contrary, in the case of *Bonamia* spp. in *C. ariakensis*, it has already been observed that the disease occurred in the warmer period when temperatures exceeded 20 °C [38]. Epidemiological circumstances of our *B. exitiosa* positive cases are unlike each of the reported cases, as no increase in mortalities or disease prevalence has been observed over the last five years.

For the reason mentioned above, the real impact of the infection is hard to evaluate. The prevalence of the *B. exitiosa* along the Croatian Adriatic Coast varied from 3.35% to a relatively high prevalence of 20% in the sampling point Brijesta in the Mali Ston Bay, and the recent recorded prevalence was 16.7% in the Medulin Bay, the site of the first finding (Table 2). These results show a higher prevalence of *B. exitiosa* compared to the findings in the study from the Manfredonia Bay in the southern part of the Italian Adriatic coast [32] in 2007, in which only two oysters out of a total of 750 were positive for *B. exitiosa*. There were no constancy in the prevalence on the sampling site over the studied period. It was shown that sequences of *B. exitiosa* detected in *O. edulis* from the Mediterranean area, including the one from Manfredonia Bay, Spain and also *O. stentina* in Tunisia [31,32,39], are 99.3% similar to those in Croatia, which supports the fact that the same strain is circulating throughout the Mediterranean basin.

Another concern is the answer to the question of how the flat oysters on Croatian production sites became infected with *B. exitiosa* It is reported that Pacific oysters are a vector or reservoirs of *Bonamia* spp. [40,41]. Although there is no farming of this species in the Croatian part of the Adriatic Sea, it has been present in the Lim Bay since 1963 [42]. This invasive species was also confirmed in the central Adriatic, but it was not found in the locality of Mali Ston, a southern production area with positive findings of *B. exitiosa* in the spring of 2016 [43]. The first positive finding of *B. exitiosa* in the Medulin Bay followed by a positive finding in Limski Bay in 2018 aroused suspicion of Pacific oysters as a source of infection for flat oysters. Unfortunately, the suspicion had to be discarded as both samples, one consisting of 18 Pacific oysters from the Medulin Bay in 2017 and another consisting of 30 Pacific oysters from the Lim Bay, tested negative for the presence of *Bonamia* spp. Furthermore, until now, there have been no records of mortality or introductions of new batches of either Pacific or flat oysters or any other mollusc species into the infected production areas. The possibility of some vector that might transfer the parasite across the Adriatic Sea should not be fully discarded, but it is hard to prove it.

Since the source of infection with *B. exitiosa* in flat oysters in all recorded production areas in Croatia remains unknown, another susceptible species may be considered a source of infection for the flat oyster. It is known that another oyster species, dwarf oysters (*O. stentina*), also exist in the Adriatic Sea [44,45], and phylogeographic research has confirmed that *B. exitiosa* has always been found in *O. stentina* [39]. The hypothesis could be made that *O. stentina*, an inhabitant in the Adriatic Sea, could be a source of *B. exitiosa*, as *O. stentina* in Tunisia were found to be positive for this parasite [39,46].

Moreover, it was not possible to get any additional information on the source of *B. exitiosa*'s origin from phylogenetic analysis. As it is visible from the phylogenetic tree, the similarity of the Croatian isolate of *B. exitiosa* to different previously sequenced isolates from Chile or Australia was 100%. The similarity of different *B. exitiosa* isolates found in different oyster species around the world emphasize the fact that the SSU rDNA gene is highly conserved. Therefore, the sequencing of additional genes or the whole genome should be carried out to provide us with more details on the phylogeny of the Croatian isolates.

More comprehensive molecular studies of the *B. exitiosa*, together with investigation of the natural population of *O. stentina* from production areas and natural beds along the Eastern Adriatic coast to confirm the natural-historical origin of the parasite *B. exitiosa*, will enable a better understanding of the pathogen's presence in the Croatian flat oyster production area.

Author Contributions: Conceptualization, D.O. and S.Z.; formal analysis, R.B., I.G.Z. and Ž.P.; investigation, R.B., Ž.P., I.G.Z., D.O. and S.Z.; resources, S.Z. and D.O.; data curation, L.M., T.M. and Ž.A.-R.; writing—original draft preparation, D.O.; writing—review and editing, S.Z., Ž.P. and R.B.; visualization, I.G.Z. and D.O.; supervision, S.Z. All authors have read and agreed to the published version of the manuscript.

Funding: This research was funded by the Croatian Ministry of Agriculture—Veterinary and Food Safety Directorate General.

Institutional Review Board Statement: Not applicable.

Informed Consent Statement: Not applicable.

Data Availability Statement: The data presented in this study are available upon a reasonable request from corresponding author.

Acknowledgments: We would like to thank Isabelle Arzul and Bruno Chollet from EURL for Molluscs Diseases for confirming the results of our findings.

Conflicts of Interest: The authors declare no conflict of interest.

References

1. MP-Uprava Ribarstva. 2019. Available online: https://ribarstvo.mps.hr/default.aspx?id=14 (accessed on 17 August 2021).
2. Grizel, H.; Mialhe, E.; Chagot, D.; Boulo, V.; Bachère, E. Bonamiosis: A model study of diseases in marine molluscs. *Am. Fish. Soc. Spec. Publ.* **1988**, *18*, 1–4.
3. Robert, R.; Borel, M.; Pichot, Y.; Gilles, T. Growth and mortality of the European oyster *Ostrea edulis* in the Bay of Arcachón (France). *Aquat. Living Resour.* **1991**, *4*, 265–274. [CrossRef]
4. Hudson, E.B.; Hill, B.J. Impact and spread of bonamiosis in the UK. *Aquaculture* **1991**, *93*, 279–285. [CrossRef]
5. Carnegie, R.B.; Cochennec-Laureau, N. Microcell parasites of oysters: Recent insights and future trends. *Aquat. Living Resour.* **2004**, *17*, 519–528. [CrossRef]
6. O'Neill, G.; Culloty, S.C.; Mulcahy, M.F. The effectiveness of two routine diagnostic techniques for the detection of the protozoan parasite *Bonamia ostreae*. *Bull. Eur. Ass. Fish. Pathol.* **1998**, *18*, 117–120.
7. Culloty, S.; Cronin, M.; Mulcahy, M. Possible limitations of diagnostic methods recommended for the detection of the protistan, *Bonamia ostreae* in the European flat oyster, *Ostrea edulis*. *Bull. Eur. Assoc. Fish. Pathol.* **2003**, *199*, 67–71.
8. Da Silva, P.M.; Villalba, A. Comparison of light microscopic techniques for the diagnosis of the infection of the European flat oyster *Ostrea edulis* by the protozoan *Bonamia ostreae*. *J. Invertebr. Pathol.* **2004**, *85*, 97–104. [CrossRef]
9. Marty, G.D.; Bower, S.M.; Clarke, K.R.; Meyer, G.; Lowe, G.; Osborn, A.L.; Chow, E.P.; Hannah, H.; Byrne, S.; Sojonky, K.; et al. Histopathology and a real-time PCR assay for detection of *Bonamia ostreae* in *Ostrea edulis* cultured in western Canada. *Aquaculture* **2006**, *261*, 33–42. [CrossRef]
10. Engelsma, M.Y.; Culloty, S.C.; Lynch, S.A.; Arzul, I.; Carnegie, R.B. *Bonamia* parasites: A rapidly changing perspective on a genus of important mollusc pathogens. *Dis. Aquat. Org.* **2014**, *110*, 5–23. [CrossRef]
11. Cochennec, N.; LeRoux, F.; Berthe, F.; Gerard, A. Detection of *Bonamia ostreae* based on small subunit ribosomal probe. *J. Invert. Pathol.* **2000**, *76*, 26–32. [CrossRef]
12. Cochennec-Laureau, N.; Reece, K.S.; Berthe, F.C.J.; Hine, P.M. *Mykrocytos roughleyi* taxonomic affiliation leads to the genus *Bonamia* (*Haplosporidia*). *Dis. Aquat. Org.* **2003**, *54*, 209–217. [CrossRef]
13. Arzul, I. Chapter 2.4.3. Infection with Bonamia ostreae. In *Manual of Diagnostic Tests for Aquatic Animals*; World Organisation for Animal Health (OIE): Paris, France, 2012.

14. Grizel, H.; Comps, M.; Raguenes, D.; Leborgne, Y.; Tigè, G.; Martin, A.G. Bilan des essais d'acclimatation d'*Ostrea chilensis* sur les côtes de Bretagne. *Rev. Trav. Inst. Pêches Marit.* **1983**, *46*, 209–225.

15. Carnegie, R.B.; Burreson, E.M.; Hine, P.M.; Stokes, N.A.; Audemard, C.; Bishop, M.J.; Peterson, C.H. *Bonamia perspora* n. sp. (*Haplosporidia*), a parasite of the oyster *Ostreola equestris*, is the first *Bonamia* species known to produce spores. *J. Eukaryot. Microbiol.* **2006**, *53*, 232–245. [CrossRef] [PubMed]

16. Carnegie, R.B.; Hill, K.M.; Stokes, N.A.; Burreson, E.M. The haplosporidian *Bonamia exitiosa* is present in Australia, but the identity of the parasite described as *Bonamia* (formerly *Mikrocytos*) *roughleyi* is uncertain. *J. Invertebr. Pathol.* **2014**, *115*, 33–40. [CrossRef]

17. Culloty, S.C.; Mulcahy, M.F. *Bonamia ostreae in the Native Oyster Ostrea edulis: A review*; Marine Environment and Health Series No. 29; Marine Institute: Rinville, Ireland, 2007; 40p.

18. Culloty, S.C.; Cronin, M.A.; Mulcahy, M.F. Potential resistance of a number of populations of the oyster *Ostrea edulis* to the parasite *Bonamia ostreae*. *Aquaculture* **2004**, *237*, 41–58. [CrossRef]

19. Hine, P.M.; Cochennec-Laureau, N.; Berthe, F.C.J. *Bonamia exitiosus* n. sp. (*Haplosporidia*) infecting flat oysters *Ostrea chilensis* in New Zealand. *Dis. Aquat. Org.* **2001**, *47*, 63–72. [CrossRef] [PubMed]

20. Burreson, E.M.; Stokes, N.A.; Carnegie, R.B.; Bishop, M.J. *Bonamia* sp. (*Haplosporidia*) found in non-native oysters *Crassostrea ariakensis* in Bogue Sound, North Carolina. *J. Aquat Anim. Health* **2004**, *16*, 1–9. [CrossRef]

21. Altschul, S.F.; Gish, W.; Miller, W.; Myers, E.W.; Lipman, D.J. Basic Local Alignment Search Tool. *J. Mol. Biol.* **1990**, *215*, 403–410. [CrossRef]

22. Ifremer, EURL-Mollusc, Edition n° 1, Bonamia ostrae and Bonamia exitiosa detection by Taqman®Real-Time Polymerase Chain Reaction. Available online: https://www.eurl-mollusc.eu/content/download/137231/file/B.ostreae%26B.exitiosa%20_TaqmanRealTimePCR_editionN%C2%B01.pdf (accessed on 17 August 2021).

23. Benson, D.A.; Cavanaugh, M.; Clark, K.; Karsch-Mizrachi, I.; Lipman, D.J.; Ostell, J.; Sayers, E.W. GenBank. *Nucleic Acids Res.* **2013**, *41*, D36–D42. [CrossRef] [PubMed]

24. Katoh, K.; Standley, D.M. MAFFT Multiple Sequence Alignment Software Version 7: Improvements in Performance and Usability. *Mol. Biol. Evol.* **2013**, *30*, 772–780. [CrossRef] [PubMed]

25. Kumar, S.; Stecher, G.; Li, M.; Knyaz, C.; Tamura, K. MEGA X: Molecular Evolutionary Genetics Analysis across Computing Platforms. *Mol. Biol. Evol.* **2018**, *35*, 1547–1549. [CrossRef]

26. Letunic, I.; Bork, P. Interactive Tree of Life (ITOL) v4: Recent Updates and New Developments. *Nucleic Acids Res.* **2019**, *47*, W256–W259. [CrossRef] [PubMed]

27. Carnegie, R.B. *Bonamiosis of Oysters Caused by Bonamia exitiosa*; Feist, S., Ed.; ICES Identification Leaflets for Diseases and Parasites of Fish and Shellfish; International Council for the Exploration of Sea: Copenhagen, Denmark, 2017; Volume 66, p. 6. [CrossRef]

28. Commission Implementing Decision (EU) 2015/1554 of 11 September 2015 Laying down Rules for the Application of Directive 2006/88/EC as Regards Requirements for Surveillance and Diagnostic Methods. Available online: https://eur-lex.europa.eu/legal-content/EN/TXT/PDF/?uri=CELEX:32015D1554&rid=1 (accessed on 17 August 2021).

29. Dinamani, P.; Hine, P.M.; Jones, J.B. Occurrence and characteristics of the haemocyte parasite *Bonamia* sp. in the New Zealand dredge oyster *Tiostrea lutaria*. *Dis. aquat. Org.* **1987**, *3*, 37–44. [CrossRef]

30. Engelsma, M.Y.; Kerkhoff, S.; Roozenburg, I.; Haenen, O.L.M.; Van Gool, A.; Sistermans, W.; Wijnhoven, S.; Hummel, H. Epidemiology of *Bonamia ostreae* infecting European flat oyster *Ostrea edulis* from Lake Grevelingen, The Netherlands. *Mar. Ecol. Prog. Ser.* **2010**, *409*, 131–142. [CrossRef]

31. Abollo, E.; Ramilo, A.; Casas, M.S.; Comesaña, P.; Cao, A.; Carballal, M.J.; Villalba, A. First detection of the protozoan parasite *Bonamia exitiosa* (*Haplosporidia*) infecting flat oyter *Ostrea edulis* grown in European waters. *Aquaculture* **2008**, *274*, 201–207. [CrossRef]

32. Narcisi, V.; Arzul, I.; Cargini, D.; Mosca, F.; Calzetta, A.; Traversa, D.; Robert, M.; Joly, J.P.; Chollet, B.; Renault., T.; et al. Detection of *Bonamia ostreae* and *B. exitiosa* (*Haplosporidia*) in *Ostrea edulis* from the Adriatic Sea (Italy). *Dis. Aquat. Org.* **2010**, *89*, 79–85. [CrossRef]

33. Carrasco, N.; Villalba, A.; Andree, K.B.; Engelsma, M.Y.; Lacuesta, B.; Ramilo, A.; Gairín, I. *Bonamia exitiosa* (*Haplosporidia*) observed infecting the European flat oyster *Ostrea edulis* cultured on the Spanish Mediterranean coast. *J. Invertebr. Pathol.* **2012**, *110*, 307–313. [CrossRef] [PubMed]

34. Buss, J.J.; Harris, J.O.; Tanner, J.E.; Wiltshire, K.H.; Deveney, M.R. Rapid transmission of *Bonamia exitiosa* by cohabitation causes mortality in *Ostrea angasi*. *J. Fish. Dis.* **2020**, *43*, 227–237. [CrossRef]

35. Lynch, S.A.; Mulcahy, M.F.; Culloty, S.C. Efficiency of diagnostic techniques for the parasite, *Bonamia ostreae*, in the flat oyster, *Ostrea edulis*. *Aquaculture* **2008**, *281*, 17–21. [CrossRef]

36. Ford, S.E.; Allam, B.; Xu, Z. Using bivalves as particle collectors with PCR detection to investigate the environmental distribution of *Haplosporidium nelsoni*. *Dis. Aquat. Org.* **2009**, *83*, 159–168. [CrossRef]

37. Arzul, I.; Gagnaire, B.; Bond, C.; Chollet, B.; Morga, B.; Ferrnad, S.; Robert, M.; Renault, T. Effects of temperature and salinity on the survival of *Bonamia ostreae* a parasite infecting flat oysters *O. edulis*. *Dis. Aquat. Org.* **2009**, *85*, 67–75. [CrossRef] [PubMed]

38. Audemard, C.; Carnegie, R.B.; Stokes, N.A.; Bishop, M.J.; Peterson, C.H.; Burreson, E.M. Effects of salinity on *Bonamia* sp. survival in Asian oyster *Crassostrea ariakensis*. *J. Shell. Res.* **2008**, *27*, 535–540. [CrossRef]

39. Hill, K.M.; Carnegie, R.B.; Aloui-Bejaoui, N.; El Gharsalli, R.; White, D.M.; Stokes, N.A.; Burreson, E.M. Observation of a *Bonamia* sp. infecting the oyster *Ostrea stentina* in Tunisia, and a consideration of its phylogenetic affinities. *J. Invert. Pathol.* **2010**, *103*, 179–185. [CrossRef] [PubMed]
40. Lynch, S.A.; Armitage, D.V.; Mulcahy, M.F.; Culloty, S.C. Investigating the possible role of benthic macroinvertebrates and zooplankton in the life cycle of the haplosporidian *Bonamia ostreae*. *Exp. Parasitol.* **2007**, *115*, 359–368. [CrossRef]
41. Lynch, S.A.; Abolle, E.; Ramilo, A.; Cao, A.; Culloty, S.C.; Villalba, A. Observations raise the question if the Pacific Oyster, *Crassostrea gigas*, can act as either a carrier or reservoir for *Bonamia ostreae* or *Bonamia exitiosa*. *Parasitology* **2010**, *137*, 1515–1526. [CrossRef]
42. Hrs-Brenko, M. *Ostrea edulis* (Linnaeus) and *Crassostrea gigas* (Thunberg) larvae in the plankton of Limski kanal in the northern Adriatic Sea. *Acta Adriat.* **1982**, *23*, 399–407.
43. Šegvić-Bubić, T.; Grubišić, L.; Zrnčić, S.; Jozić, S.; Žužul, I.; Talijančić, I.; Oraić, D.; Relić, M.; Katavić, I. Range expansion of the non-native oyster. *Acta Adriat.* **2016**, *57*, 321–330.
44. Hrs-Brenko, M.; Legac, M. Inter-and intra-species relationships of sessile bivalves on the eastern coast of the Adriatic Sea. *Nat. Croat.* **2006**, *15*, 203–230.
45. Vondrak, T. Upotreba Molekularnih Biljega u Identifikaciji Različitih vrsta Kamenica u Jadranu, Diplomski rad, Sveučilište u Zagrebu, Prirodoslovno-Matematički fakultet, Biološki Odsjek. 2017. Available online: https://repozitorij.pmf.unizg.hr/islandora/object/pmf%3A4522/datastream/PDF/view (accessed on 17 August 2021).
46. Hill-Spanik, K.M.; McDowell, J.R.; Stokes, N.A.; Reece, K.S.; Burreson, E.M.; Carnegie, R.B. Phylogeographic perspective on the distribution and dispersal of a marine pathogen, the oyster parasite *Bonamia exitiosa*. *Mar. Ecol. Prog. Ser.* **2015**, *536*, 65–76. [CrossRef]

Journal of
Marine Science and Engineering

Article

Histopathologic Lesions in Bivalve Mollusks Found in Portugal: Etiology and Risk Factors

Daniel Pires [1], Ana Grade [2], Francisco Ruano [2] and Fernando Afonso [1,*]

1   CIISA—Centre for Interdisciplinary Research in Animal Health, Faculty of Veterinary Medicine, University of Lisbon, Av. da Universidade Técnica, 1300-477 Lisbon, Portugal; dannigs10@hotmail.com
2   IPMA—Instituto Português do Mar e da Atmosfera, Av. Alfredo Magalhães Ramalho, 1495-165 Lisboa, Portugal; agrade@ipma.pt (A.G.); fruano@ipma.pt (F.R.)
*   Correspondence: fafonso@fmv.ulisboa.pt

Abstract: Bivalve mollusks are an important resource due to their socioeconomic value and to the historical and genetic value of some species. Two nationally important oyster species-Portuguese oyster (*Crassostrea angulata*) and Japanese oyster (*Crassostrea gigas*) from distinctive areas in Portugal were studied to evaluate their sanitary status. Oysters were sampled from four different sites in Portugal. Oysters collected from Japanese oyster populations were cultivated in a strong ocean-influenced environment and Portuguese oyster populations were cultivated in wild-beds. The histopathological examination of both oyster species revealed the presence of parasites in gills, mantle epithelium, digestive gland tubules and connective tissue, with a moderate prevalence. In both populations was observed hemocytosis in the connective tissue, edema and metaplasia in the digestive gland and tissues necrosis. In wild populations from Sado and Mira estuaries the prevalence of mud blisters and gill lesions were higher than from populations produced on 0.50 m tables from mudflats. Biosecurity measures and diagnostic techniques are fundamental to control pathogenic agents, including the identification of pathogens at an early stage in their life cycles. This will prevent diseases and improve pathogen reduction on transport of animals from different countries and regions to new production areas to avoid the transmission of diseases.

Keywords: bivalve mollusks; oysters; histopathology; parasites

Citation: Pires, D.; Grade, A.; Ruano, F.; Afonso, F. Histopathologic Lesions in Bivalve Mollusks Found in Portugal: Etiology and Risk Factors. *J. Mar. Sci. Eng.* **2022**, *10*, 133. https://doi.org/10.3390/jmse10020133

Academic Editor: Snježana Zrnčić

Received: 1 October 2021
Accepted: 30 December 2021
Published: 20 January 2022

Publisher's Note: MDPI stays neutral with regard to jurisdictional claims in published maps and institutional affiliations.

1. Introduction

In Portugal, in the late 20th century, the two most important commercial oysters were the flat Oyster (*Ostrea edulis*) and the Portuguese Oyster (*Crassostrea angulata*) [1]. Initially, flat oyster commerce was more important, but in the 20th century it was replaced by the Portuguese oyster. The culture of bivalve mollusks is an activity with high expression in Portuguese aquaculture. It represents 55% of the whole production, being the main species, clams, oysters and mussels [2].

There was a significant change in bivalve mollusks production. Initially, it was exclusively chosen for semi-intensive production in large areas. Nevertheless, at this moment there is an intensification of production in less area, increasing bivalve mollusks density. This option results in lower growth rates, lower product quality, lower fertility rates and increasing diseases [3].

Microbiological contamination of water where aquaculture production is present, in most cases, is the result of industries, urban activities and leisure activities, causing fecal contamination. [4–6]. The microbiological organisms present in bivalve mollusks are diverse, with different populations of bacteria (namely, *Escherichia coli*, *Clostridium perfringens*, *Vibrio parahaemolyticus*, *Salmonella* and *Listeria*) [7–9], enteric viruses (namely, norovirus, enteric calicvirus, hepatitis A virus and other enteroviruses) [10,11] and protozoa (namely, *Cryptosporidia* and *Giardia*) [12,13]. Generally, these microorganisms are not harmful to bivalve mollusks and do not cause lesions or disease.

Several diseases can affect different bivalve mollusks species and their production. Epizootics caused by fungi, viruses and protozoa can affect the entire bivalve production, as in Portugal and France in the 1970s where massive mortality caused by an iridovirus affected the Portuguese oyster production. Since 2009 a herpes virus (OsHV-1) has been the cause of massive mortalities in Japanese oyster (*Crassostrea gigas*), first in France and all over the most important European production countries of this species, including Portugal [3]. Japanese oyster aquaculture is a major primary production sector in many countries, but the industry has been threatened by mortality events for the last five decades [14]. In Portuguese and Japanese oyster, the main pathogens are virus, namely iridovirus, responsible for gill disease that affected Portuguese oyster production in the 1970's, and herpes viruses, namely Ostreid-herpesvirus 1 μvar (OsHV-1) that is related with summer oyster disease and affects the Japanese and Portuguese oyster populations [15]. From the first detection in the Eastern oysters (*Crassostrea virginica*) [14] in 1972, herpes virus has been identified in at least 20 bivalve species and has caused massive mortalities [14]. Herpes virus infecting bivalve mollusks are virulent pathogens of both larval and seed oysters. Mortalities of juvenile oysters associated with detection of herpes viruses have been reported from France, New Zealand, Spain and the USA [16–19].

In flat oyster, there are important diseases, including bonamiosis that is caused by *Bonamia ostrea* and *Marteilia refringens* that causes marteliosis and viruses such as *Ostreid-herpesvirus* 1 μvar [20,21].

Regarding mussels *Mytilus* sp., the main pathogens are protozoa, such as *Marteilia refringens* that is the causative agent of marteliosis and copepods, such as *Mytilicola intestinalis* which causes red worm disease [22,23].

Japanese clam, European clam and cockles have been mainly affected by protozoa, such as *Perkinsus olseni*, the causative agent of Perkinsiosis [24–27] and *Haplosporidium tapetis* [26,28]. *Perkinsus olseni* is responsible for the high mortality of cockleshell populations. Perkinsiosis is caused by the protozoan *P. olseni*. This disease is a very common disease in clams and has caused massive mortalities in clam populations, contributing to the decrease in their production and consequently affecting their socioeconomic value [29]. Typical lesions that occur include cell disorganization, autolysis and necrosis of cells, both within and near the lesion [24].

The high prevalence of this pathogen in clams can be attributed to several factors, such as degraded sediments, high animal densities per unit area, lack of physical barriers between different culture beds, and transfer of animals carrying the disease [1].

Biotoxins can affect the nervous, the gastrointestinal and the respiratory systems, being fatal in some severe cases. They originate in various microalgae and dinoflagellates that produce these toxic elements in their metabolism. Mollusk bivalves are filter-feeder animals, and they can retain and accumulate those microalgae and the biotoxins, and as such, they can be a source of contamination for humans [30,31].

The accumulation of biotoxins in aquatic organisms depends on their feeding activity, on their metabolism, and on their elimination rate, modifying the transfer of toxins in the trophic chain [32].

Contaminant metals can cause several problems in humans, and the contamination of bivalve mollusks with different concentrations of heavy metals can be dangerous, namely lead, cadmium and mercury. They can accumulate high concentrations from water, and from the sediments and they have a wide bioaccumulation range of heavy metals depending on the species [33–35].

The aim of the present work was to study the sanitary condition in oyster production in four different populations of oysters. The objective was to describe the general health status, studying anatomo and histopathological lesions and parasites, in the Portuguese oyster (*Crassostrea angulata*) and Japanese oyster (*Crassostrea gigas*) from different production areas in Portugal, in order to evaluate their health condition.

2. Materials and Methods

The map (Figure 1) shows the four different areas where oysters were collected.

Figure 1. The map shows the areas were oysters were collected in mainland Portugal: Aveiro lagoon (1), Sado estuary (2), Mira estuary (3) and Alvor lagoon (4).

Oysters (9–11 cm, aged 24 months old) were sampled from four different sites of the Portuguese coast. *Crassostrea gigas* is an exotic species that was introduced in Portugal, and so, its presence is restricted to some production areas (1 and 4). In areas 2 and 4, *Crassostrea angulata* were collected from wild beds.

Portuguese oysters from Sado estuary ($n = 30$) and Mira estuary ($n = 30$) and Japanese oyster from Alvor lagoon ($n = 30$) and Aveiro lagoon ($n = 30$) were collected in May 2019. In May, in the four sites, oysters have already spawned and are more susceptible to environment stressors and increasing chances of histopathological lesions.

Individuals were randomized collected in production areas present in four mesotidal estuarine areas indicated in the map (Figure 1). The areas are located on the west Portuguese coast (1, 2 and 3) and in the south Portuguese coast (4). All the four production areas are under a strong oceanic influence because of the proximity to the coastal water and they have an influence of fresh water. For production, oysters are placed inside plastic bags that are put on tables with a distance of 0.50 m from the bottom in areas 1 and 4. In Sado and Mira estuaries, oysters were collected in intertidal mudflats where oyster beds are located. In Aveiro lagoon there is a lagoon with several channels, stretching for 45 Km. Sado stuary includes a section of river, marshlands and channels and the dynamic tide extends for 65 Km. Mira estuary is a narrow estuary of the "Ria" pattern that extends for 40 km. Alvor lagoon has a water body and a mesotidal shallow lagoon that extends for 15 Km.

Oyster were collected during low tide at a distance from the mouth of 1 Km (1), 17 Km (2), 15 Km (3) and 1 Km (4). Temperature and salinity of the water were, respectively: (1) 18.6 and 31.5; (2) 19.1 and 33.1; (3) 19.7 and 33.6; (4) 20.1 and 31.9.

After being collected, oysters were immediately transported on ice to the laboratory. To survey the presence of lesions, parasites and diseases anatomic and histopathological examinations were used as main diagnostic methods. Oysters were opened with a knife and the samples, including all organs and tissues, were collected with a tweezer and a scalpel, Tissue samples were prepared for histopathology processing, following the protocol used in Pathology laboratory of IPMA. Samples were fixed in Davidson's fixative for 48 h, dehydrated and embedded in paraffin. Sections with 5 µm thick were stained with Hematoxylin-Eosin and mounted on a microscopic glass slide.

Histological preparations were carefully examined under light microscopy (Motic BA-410), looking for the presence of lesions and parasites in oysters.

3. Results

In most of the samples no lesions were observed in organs (Figure 2A–C) The macroscopic lesions (Figure 3A–C), microscopic lesions (Figure 4A–F,I) and parasites (Figure 4G,H) were recorded.

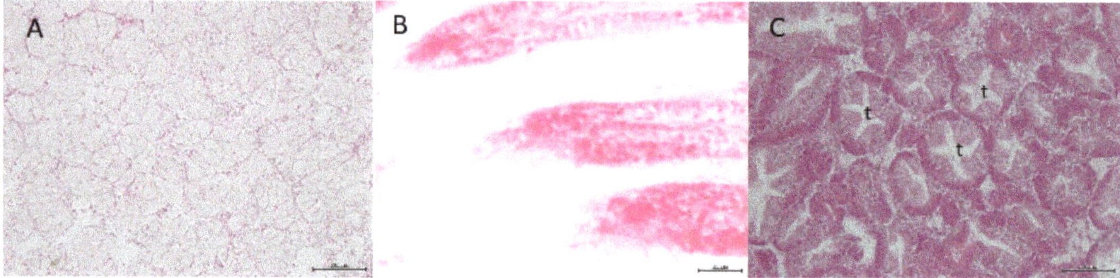

Figure 2. Histological sections stained with hematoxylin and eosin, where different tissues and organs are shown. (**A**) Connective tissue normal is present (Mag = 40×; bar = 200 µm) (*C. angulata* tissue). (**B**) The normal gill with filaments where a ciliary system is observed (*C. gigas* tissue) (Mag = 200×; bar = 20 µm). (**C**) The normal digestive gland presents a regular tubule (t) thickness, characterized by their dense areas and star shape (*C. gigas* tissue). (Mag = 40×; bar = 200 µm).

Figure 3. Macroscopical lesions observed in oysters. (**A**) Gills lesions (arrow) in *C. angulata* (bar = 3 cm). (**B**) Shell disease (arrow) in *C. angulata* (bar = 3 cm). (**C**) Mud blisters (arrows) in *C. gigas* (bar = 3 cm).

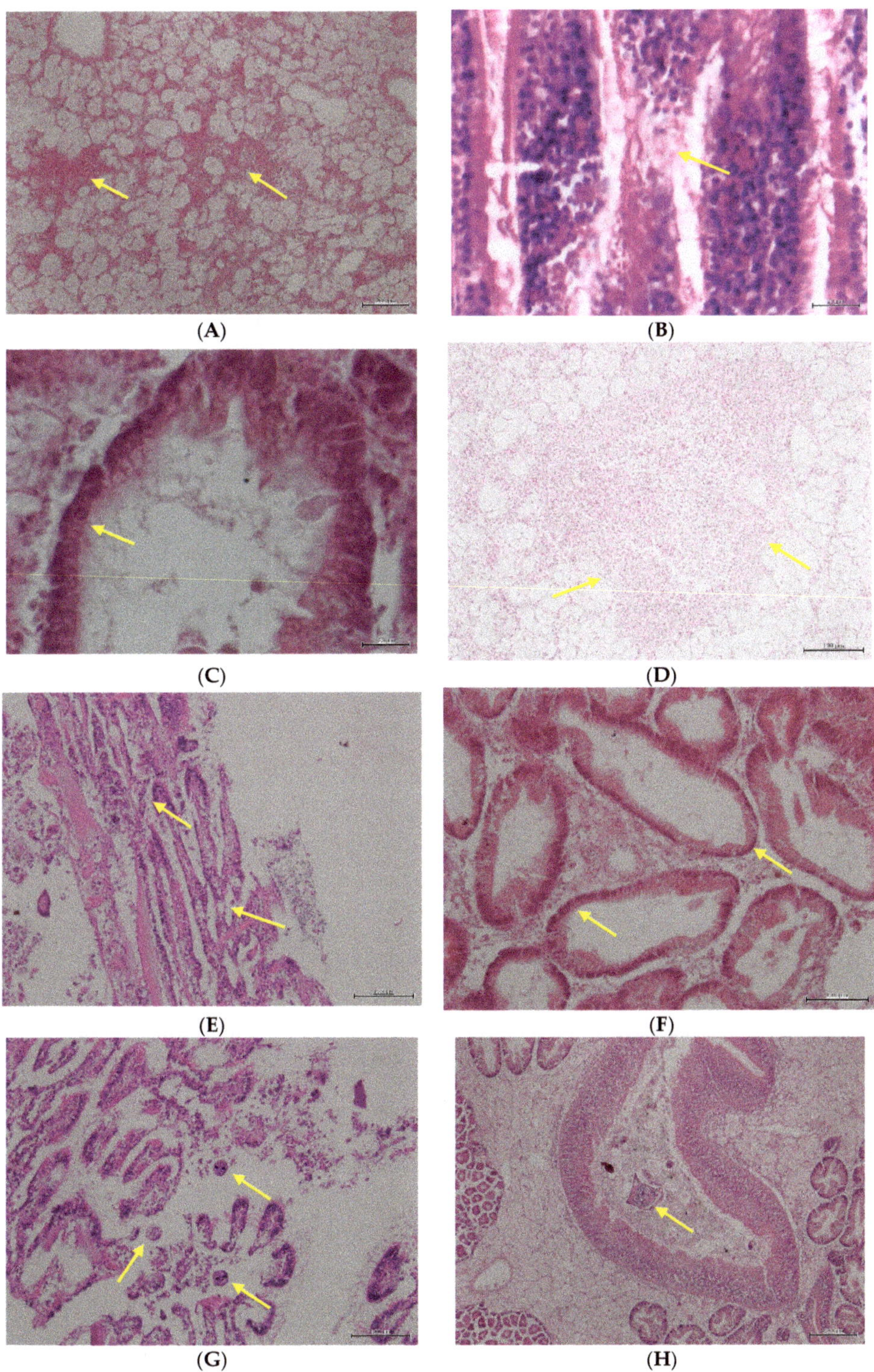

Figure 4. *Cont.*

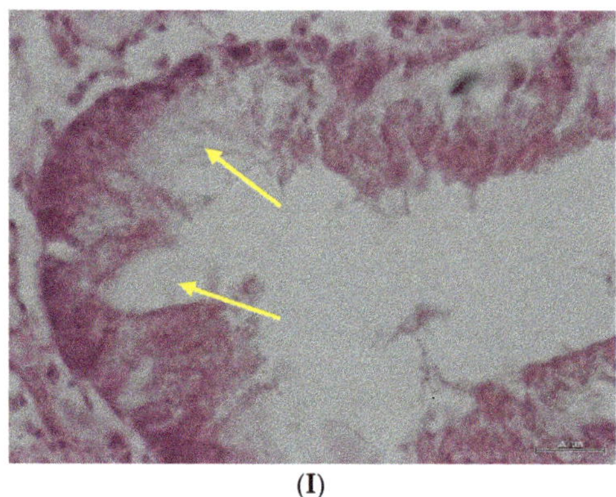

(I)

Figure 4. Microscopical lesions observed in oysters (tissue sections are stained with hematoxylin and eosin). (**A**,**D**) An inflammatory response is observed with the presence of hemocyte in the connective tissue (arrows) (*C. gigas* tissue). (Mag = 40×; bar = 200 μm). (**B**) Necrosis of gill tissue (arrow) (*C. angulata* tissue) (Mag = 200×; bar = 20 μm). (**C**) Metaplasia of digestive gland (arrows) (*C. angulata* tissue). (Mag = 200×; bar = 20 μm). (**E**) Necrosis of gill tissue (arrows) (*C. angulata* tissue). (Mag = 100×; bar = 100 μm). (**F**) Metaplasia of digestive gland (arrows) (*C. angulata* tissue). (Mag = 100×; bar = 100 μm). (**G**) *Trichodina* sp. (arrows) in gill tissue (*C. angulata* tissue). (Mag = 100×; bar = 100 μm). (**H**) *Mytilicola* sp. (arrow) in intestine (*C. gigas* tissue). (Mag = 40×; bar = 200 μm). (**I**) Loss of epithelium of the digestive gland (arrows) (*C. gigas* tissue) (Mag = 200×; bar = 20 μm).

In samples of *C. gigas* from Aveiro lagoon, it was observed only 3% of oysters with edema, 3% with shell disease and 10% with *M. ostrea*.

Finally, in samples from Aveiro lagoon, a prevalence of 13% *Trichodina* sp. infections, 13% *Ancistrocoma* sp. infections and 10% *Mytilicola* sp. infections. Concerning tissue lesions and morphological changes, 26% of wild oysters showed necrosis, 45% showed hemocytosis, indicating an inflammatory process, mainly in connective tissue. It was also observed a prevalence of 3% of ceroidosis in the connective tissue, edema in 15 % and 18% showed metaplasia in the digestive gland. Regarding farmed oysters, the following lesions and prevalence were found: necrosis (6%) in different tissues, hemocytosis (27%), ceroidosis (5%), edema (18%) and metaplasia (5%).

Samples of wild *C. angulata* from Sado estuary show the following results (Tables 1 and 2): 20% presented mud blisters, mainly caused by *Polydora* sp.; 33% with gills lesions; 10% had shell disease associated with the presence of the fungus *Ostracoblabe implexa*; and 17% showed *Myicola ostrea* copepods.

Table 1. Prevalence of macroscopic lesions and parasites in different populations.

Macroscopic **Lesions and Parasites**	Site 1	Site 2	Site 3	Site 4
Tricodina sp.	13%	0%	13%	23%
Ancistrocoma sp.	13%	0%	30%	13%
Mytilicola sp.	10%	0%	3%	10%
Myicola ostrea	10%	17%	0%	0%
Shell disease	0%	10%	10%	0%
Mud blisters	3%	20%	97%	10%
Gills lesions	0%	33%	50%	0%

Table 2. Prevalence of microscopic lesions in different populations.

Microscopic Lesions	Site 1	Site 2	Site 3	Site 4
Necrosis	13%	13%	40%	23%
Hemocytosis	36%	36%	53%	16%
Ceroidosis	10%	0%	6%	0%
Edema	36%	0%	30%	0%
Metaplasia	3%	26%	10%	6%

Portuguese oysters sampled in Sado estuary, it was found a prevalence of 30% infections with *Trichodina* sp., characterized by a shape similar to a disc, a circlet of eosinophilic denticles, a ciliary fringe and a horseshoe shaped nucleous, in gills and mantle epithelium. A prevalence of 20% infections with *Ancistrocoma* sp., characterized by a Spindle-shaped ciliates with large, granular and polymorphic nuclei, in the digestive gland tubules and connective tissue. Copepods (*Mytilicola* sp.), characterized by an elongated cylindrical shape and a redish colour were observed within the intestine of oysters with the prevalence of 17% [36].

Samples of wild *C. angulata* Mira estuary showed mud blisters (96%), 50% had gills lesions, 10% presented edema and 10% shell disease.

In samples from Mira estuary, it was found a prevalence of 13% infections with *Trichodina* sp., 30% infections with *Ancistrocoma* sp. and 3% infections with *Mytilicola* sp.

Finally, in samples of *C. gigas* from Alvor lagoon 10% presented mud blisters.

In farmed oysters sampled from Alvor lagoon, it was observed a prevalence of 23% *Trichodina* sp. infections and a prevalence of 27% *Ancistrocoma* sp. Copepod (*Mytilicola* sp.) were observed within the intestine of oysters at a prevalence of 3%.

4. Discussion

Cultivation and harvesting of bivalve mollusks are two important activities, and it is fundamental to preserve the natural ecosystems that support those activities and lead to increasing natural production and improving its quality. Bivalve mollusks production depends on the external factors [37], including water parameters, such as temperature variations, food change, biotoxins and pathogens (including parasites, bacteria, viruses) or anthropogenic, such as chemical contaminants (heavy metals, pesticides), organic contaminants (fecal contamination) and overexploitation of natural banks [1,38]. It is wildly known that the marine environment provides stress to oysters during their life cycle, namely changes in environmental parameters, changes in the availability of food and the display of various toxic pollutants [39–41]. Oysters are very resistant to changes of temperature and salinity that are present in the ecosystems where they live, namely estuaries and lagoons. As other euryhaline organisms, oysters can live in water where large variations of salinity occur. For short periods of time salinity values could range between 2% and 38% [1,42]. Similarly, oysters are resistant to temperature changes between 8 °C and 30 °C [1]. There are several factors that influence the growth of mollusk bivalves such as temperature, oxygen and the presence of different species of phytoplankton. Once they are filter-feeders, the quality and the quantity of filtered food define the outcome of their growth [43].

The regularly consumption of bivalve mollusks by different populations and the growth of shellfish aquaculture production show the importance of these animals and their monitoring must be undertaken to avoid diseases and mortality in oyster populations. Hereby we are presenting the results of the sanitary monitoring of oyster production in four different production areas. In general, it was observed that low numbers of parasites were present in oysters being *Trichodina* sp. and *Ancistrocoma* sp. infections the most common ones present in the areas that were studied. The presence of *Trichodina* sp., *Ancistrocoma* sp., *Mytilicola* sp. and *Myicola ostrea* are commonly observed in oyster populations and in massive infestations could lead to host weakness [36,44–46]. Lesions of the internal organs, such as metaplasia, may reflect physiological stress, contaminants or the presence of large parasitic loads [1,44]. In our study, hemocytosis, cerodosis and edema were mainly present in the interstitial connective tissue. These lesions and those observed in the epithelium of the diverticula of the digestive gland, including the hemocytic infiltration and necrosis were usually associated and could be related with the presence of the parasites. In wild populations from Sado and Mira estuaries the prevalence of mud blisters and gill lesions were higher than from populations produced on tables 0.50 m from mudflats.

The identification, characterization and registration of pathologic processes in oysters constitute a set of important measures for sanitary control. Based on the results of the sanitary control appropriate biosecurity tools could be developed and implemented leading to the control of diseases spread and transmission. Finally, knowledge and ability to manage risk of different pathogen will contribute to the sustainability of oyster production and protection of animal and human health.

Aquatic ecosystems can be modified due to human activities such as overharvesting and destruction of substrates [47]. In addition, transfers of oysters, between countries and regions can spread diseases and invasive species [47–49]. Concurrently with increase of the socio-economic importance of mollusk farming in Portugal all associated risk factors are affecting this production. It is essential to have stricter control at all stages of production of the different bivalve mollusk species. This is the only way to improve economic performance without jeopardizing animal welfare and at the same time to prevent diseases in oyster populations and their dissemination to different areas. In Portugal, IPMA (https://www.ipma.pt/en/) (accessed on 14 September 2021) undertakes regular analysis to the water, including in the areas where animal species that are produced to be consumed by populations. In the areas where animals were collected no contaminants were detected. Biotoxins are not harmful to bivalve mollusks but, when they are consumed by humans, they can cause several serious problems. As such, the control and monitoring of levels of contaminants are fundamental to prevent any risk to public health.

Diseases are important risk factors, specific or not, of different species of bivalve mollusks, which are caused, in many cases, by non-compliance with basic management rules, namely the level of animal load in production areas, the length of time the bivalve mollusks remain in these areas, and the introduction of seeds of unknown origin. Effective biosecurity measures and correct diagnostic techniques are essential to control pathogenic agents. The identification of pathogens should include those at an early stage in their life cycles, to control and to prevent their multiplication and proliferation. In the production areas from Aveiro and Alvor lagoons these measures are implemented by producers and by the authorities. Producers understand that it is essential to prevent mortalities in bivalve mollusk populations, namely avoiding overcrowding and preventing diseases. In Sado and Mira estuaries, wild populations show that oysters are more susceptible to lesions than oysters produced on tables under the supervision of producers.

Research on genetic resistant bivalve populations to different pathogenic agents and the use of probiotic bacteria are important strategies to prevent diseases caused, namely by different *Vibrio* species [50,51] that can cause mortality outbreaks, that did not occur in the areas that were studied. Furthermore, restoration programs can be implemented in areas where ecosystems were negatively affected [49]. The success of bivalve mollusk production can only be achieved, if appropriate measures are applied, namely biosecurity, correct and efficient diagnostic methods, effective compliance with legislation and the dissemination of knowledge among populations.

Author Contributions: Methodology (F.A.; F.R.); Investigation (A.G.; D.P.); Writing—original draft preparation (D.P.; F.A.); Writing—review and editing (A.G.; F.R.). All authors have read and agreed to the published version of the manuscript.

Funding: This work was supported by MAR2020: MAR-02.05.01-FEAMP-0010.

Institutional Review Board Statement: Not applicable.

Informed Consent Statement: Not applicable.

Data Availability Statement: Not applicable.

Conflicts of Interest: The authors declare no conflict of interest.

References

1. Ruano, F. Fisheries and farming of important marine bivalves in Portugal. *NOAA Tech. Rep. NMFS* **1997**, *3*, 191–200.
2. Government of Portugal. *Fisheries Statistics 2018*; Government of Portugal: Lisboa, Portugal, 2019.
3. Grizel, H.; Bachere, E.; Mialhe, E.; Tige, G. Solving parasite-related problems in cultured molluscs. *Int. J. Parasit.* **1987**, *17*, 301–308. [CrossRef]
4. McAllister, T.; Topp, E. Role of livestock in microbiological contamination of water: Commonly the blame, but not always the source. *Anim. Front.* **2021**, *2*, 17–27. [CrossRef]
5. D'Ugo, E.; Marcheggiani, S.; D'Angelo, A.; Caciolli, S.; Puccinelli, C.; Giuseppetti, R.; Marcoaldi, R.; Romanelli, C.; Mancini, L. Microbiological water quality in the medical device industry in Italy. *Microchem. J.* **2018**, *136*, 293–299. [CrossRef]
6. Maran, N.; Crispim, B.; Iahnn, S.; Araújo, R.; Grisolia, A.; Oliveira, K. Depth and Well Type Related to Groundwater Microbiological Contamination. *Int. J. Environ. Res. Public Health* **2016**, *13*, 1036. [CrossRef] [PubMed]
7. Pasquale, V.; Romano, V.; Rupnik, M.; Capuano Bove, D.; Aliberti, F.; Krovacek, K.; Dumontet, S. Occurrence of toxigenic Clostridium difficile in edible bivalve molluscs. *Food Microbiol.* **2012**, *31*, 309–312. [CrossRef] [PubMed]
8. Martinez, O.; Rodriguez-Calleja, J.; Santos, J.; Otero, A.; Garcia-Lopez, M.L. Foodborne and Indicator Bacteria in Farmed Molluscan Shellfish before and after Depuration. *J. Food Prot.* **2009**, *72*, 1443–1449. [CrossRef] [PubMed]
9. Ukwo, S.; Ezeama, C.; Obot, O. Microbiological safety and toxic element contaminants in bivalve shellfish from intertidal mudflats of IKO estuary, Niger delta, Nigeria. *South Asian J. Food Technol. Environ.* **2019**, *5*, 846–854. [CrossRef]
10. Gabrieli, R.; Macaluso, A.; Lanni, L.; Saccares, S.; Giamberardino, F.; Petrinca, B.; Divizia, M. Enteric viruses in molluscan shellfish. *New Microbiol.* **2007**, *30*, 471–475.
11. Iaconelli, M.; Purpari, G.; Della Libera, S.; Petricca, S.; Guercio, A.; Ciccaglione, A.; Bruni, R.; Taffon, S.; Equestre, M.; Fratini, M.; et al. Hepatitis A and E Viruses in Wastewaters, in River Waters and in Bivalve Molluscs in Italy. *Food Environ. Virol.* **2015**, *7*, 316–324. [CrossRef]
12. Schets, F.; Van Den Berg, H.; Husman, A. Determination of the Recovery Efficiency of Cryptosporidium Oocysts and Giardia Cysts from Seeded Bivalve Mollusks. *J. Food Prot.* **2013**, *76*, 93–98. [CrossRef] [PubMed]
13. Gomez-Couso, H.; Mendez-Hermida, F.; Castro-Hermida, J.; Ares-Mazás, E. Cryptosporidium Contamination in Harvesting Areas of Bivalve Molluscs. *J. Food Prot.* **2006**, *69*, 185–190. [CrossRef] [PubMed]
14. Alfaro, A.; Nguyen, T.; Merien, F. The complex interactions of Ostreid herpesvirus 1, Vibrio bacteria, environment and host factors in mass mortality outbreaks of Crassostrea gigas. *Aquaculture* **2019**, *11*, 1148–1168. [CrossRef]
15. Barbosa-Solomieu, V.; Renault, T.; Travers, M.-A. Mass mortality in bivalves and the intricate case of the Pacific oyster, Crassostrea gigas. *J. Invertebr. Pathol.* **2015**, *131*, 2–10. [CrossRef] [PubMed]
16. Burge, C.; Judah, L.R.; Conquest, L.; Griffin, F.; Cheney, D.; Suhrbier, A.; Vadopalas, B.; Olin, P.; Renault, T.; Friedman, C. Summer seed mortality of the Pacific oyster, Crassostrea gigas Thunberg grown in Tomales Bay, California, USA: The influence of oyster stock, planting time, pathogens, and environmental stressors. *J. Shellfish Res.* **2007**, *26*, 163–172. [CrossRef]
17. Hine, P.; Wesney, B.; Hay, B. Herpesviruses associated with mortalities among hatchery-reared Pacific oysters, Crassostrea gigas. *Dis. Aquat. Org.* **1992**, *12*, 135–142. [CrossRef]
18. Renault, T.; Le Deuff, R.; Lipart, C.; Delsert, C. Development of a PCR procedure for the detection of a herpes-like virus infecting oysters in France. *J. Virol. Meth.* **2000**, *88*, 41–50. [CrossRef]
19. Elandaloussi, L.; Carrasco, N.; Andree, K.; Furones, D.; Roque, A. Esdeveniments de mortalitat de lostró del Pacific (*Crassostrea gigas*) en el delta del Ebre. Estudi de cas. In Proceedings of the II Simposi d'aqüicultura de Catalunya, Sant Carles de la Rapita, Spain, 15–17 October 2009.
20. World Organisation for Animal Health (OIE). Infection with *Bonamia ostreae*. In *Manual of Diagnostic Tests for Aquatic Animals*; World Organisation for Animal Health: Paris, France, 2018; Chapter 2.4.3.
21. The Food and Agriculture Organization (FAO). Cultured Aquatic Species Information Programme. *Ostrea edulis* (Linnaeus, 1758). 2018. Available online: http://www.fao.org/fishery/culturedspecies/Ostrea_edulis/en (accessed on 15 May 2021).
22. The Food and Agriculture Organization (FAO). Cultured Aquatic Species Information Programme. *Mytilus galloprovincialis* (Lamarck, 1819). 2018. Available online: http://www.fao.org/fishery/culturedspecies/Mytilus_galloprovincialis/en (accessed on 15 May 2021).
23. The Food and Agriculture Organization (FAO). Cultured Aquatic Species Information Programme. *Mytilus edulis* (Linnaeus, 1758). Available online: http://www.fao.org/fishery/culturedspecies/Mytilus_edulis/en (accessed on 15 May 2021).
24. Azevedo, C. Fine structure of *Perkinsus atlanticus* sp. (Apicomplexa, Perkinsea) parasite of the clam Ruditapes decussatus from Port. *J. Parasitol.* **1989**, *75*, 627–635.
25. Azevedo, C. Virus-like particles in *Perkinsus atlanticus* (Apicomplexa, Perkinsidae). *Dis. Aquat. Org.* **1990**, *9*, 63–65. [CrossRef]
26. The Food and Agriculture Organization (FAO). Cultured Aquatic Species Information Programme. *Ruditapes decussatus* (Linnaeus, 1758). 2018. Available online: http://www.fao.org/fishery/culturedspecies/Ruditapes_decussatus/en (accessed on 15 May 2021).

27. The Food and Agriculture Organization (FAO). Cultured Aquatic Species Information Programme. *Ruditapes philippinarum* (Adams & Reeve, 1850). 2018. Available online: http://www.fao.org/fishery/culturedspecies/Ruditapes_philippinarum/en (accessed on 15 May 2021).

28. Vilela, H. Sporozoaires parasites de la palourde, *Tapes decussatus* (L.). In *Revista da Faculdade de Ciências, Lisboa*; University of Lisbon: Lisbon, Portugal, 1951; Volume 1, pp. 379–386.

29. Ruano, F.; Batista, F.; Arcangeli, G. Perkinsosis in the clams *Ruditapes decussatus* and *Ruditapes philippinarum* in the Northeastern Atlantic and Mediterranean Sea: A review. *J. Invertebr. Pathol.* **2015**, *131*, 58–67. [CrossRef]

30. Vale, P. *Marine Biotoxins*; Faculty of Veterinary Medicine, Technical University of Lisbon: Lisbon, Portugal, 2004; Volume 99, pp. 3–18.

31. Bricelj, V.; Shumway, S. Paralytic shellfish toxins in bivalve Molluscs: Occurence, transfer kinetics, and biotransformation. *Rev. Fish. Sci.* **1998**, *6*, 315–383. [CrossRef]

32. Costa, P.; Costa, S.; Braga, A.; Rodrigues, S.; Vale, P. Relevance and challenges in monitoring marine biotoxins in non-bivalve vectors. *Food Control* **2017**, *76*, 24–33. [CrossRef]

33. Gupta, S.; Singh, J. Evaluation of mollusc as sensitive indicatior of heavy metal pollution in aquatic system: A review. *IOAB J.* **2011**, *2*, 49–57.

34. Nour, H. Distribution and accumulation ability of heavy metals in bivalve shells and associated sediment from Red Sea coast, Egypt. *Environ. Monit. Assess* **2020**, *192*, 353. [CrossRef]

35. Souza, R.; Garbossa, L.; Campos, C.; Vianna, L.; Vanz, A.; Rupp, G. Metals and pesticides in commercial bivalve mollusc production areas in the North and South Bays, Santa Catarina (Brazil). *Mar. Pollut. Bull.* **2016**, *105*, 377–384. [CrossRef] [PubMed]

36. Bower, S.; Mcgladdery, S.; Price, I. Synopsis of Infectious Diseases and Parasites of Commercially Exploited Shellfish. *Annu. Rev. Fish Dis.* **1994**, *4*, 1–199. [CrossRef]

37. Gallardi, D.; Dörner, J.; Carbonell, P.; Pino, S.; Farías, A. Effects of Bivalve Aquaculture on the Environment and Their Possible Mitigation: A Review. *Fish. Aquac. J.* **2014**, *5*, 3. [CrossRef]

38. Wan, W.; Lu, G. Heavy Metals in Bivalve Mollusks. In *Chemical Contaminants and Residues in Food*, 2nd ed.; Schrenk, D., Cartus, A., Eds.; Woodhead Publishing Limited: Cambridge, UK, 2017; pp. 553–594.

39. Lacoste, A.; Jalabert, F.; Malham, S.K.; Cueff, A.; Poulet, S.A. Stress and Stress-Induced Neuroendocrine Changes Increase the Susceptibility of Juvenile Oysters (*Crassostrea gigas*) to Vibrio splendidus. *Appl. Environ. Microbiol.* **2001**, *67*, 2304–2309. [CrossRef]

40. Kennedy, V.; Newell, R.; Eble, F. *The Eastern Oyster Crassostrea virginica*; Maryland Sea Grant College: College Park, MD, USA, 1996.

41. Garnier, M.; Labreuche, Y.; Garcia, C.; Robert, M.; Nicolas, J.-L. Evidence for the Involvement of Pathogenic Bacteria in Summer Mortalities of the Pacific Oyster Crassostrea gigas. *Microb. Ecol.* **2007**, *53*, 187–196. [CrossRef] [PubMed]

42. Héral, M.; Deslous-Paoli, J. Oyster Culture in European Countries. In *Estuarine and Marine Bivalve Mollusk Culture*; Menzel, W., Ed.; CRC Press: Boca Raton, FL, USA, 1991; pp. 153–190.

43. Sarà, G.; Mazzola, A. Effects of trophic and environmental conditions on the growth of *Crassostrea gigas* in culture. *Aquaculture* **1997**, *153*, 81–91. [CrossRef]

44. Iglesias, D.; Rodriguez, L.; Gómez, L.; Azevedo, C.; Montes, J. Histological survey of Pacific oysters Crassostrea gigas (Thunberg) in Galicia (NW Spain). *J. Invertebr. Pathol.* **2012**, *111*, 244–251. [CrossRef]

45. Sabry, R.; Gesteira, T.; Magalhães, A.; Barracco, M.; Guertler, C.; Ferreira, L.; Vianna, R.; Silva, P. Parasitological survey of mangrove oyster, Crassostrea rhizophorae, in the Pacoti River Estuary, Ceará State, Brazil. *J. Invertebr. Pathol.* **2013**, *112*, 24–32. [CrossRef] [PubMed]

46. Aguirre-Macedo, M.; Kennedy, C. Diversity of metazoan parasites of the introduced oyster species Crassostrea gigas in the Exe Estuary. *J. Mar. Biol. Assoc. U.K.* **1999**, *79*, 57–63. [CrossRef]

47. Pogoda, B. Current Status of European Oyster Decline and Restoration in Germany. *Humanities* **2019**, *8*, 9. [CrossRef]

48. Bromley, C.; McGonigle, C.; Ashton, E.C.; Roberts, D. Bad moves: Pros and cons of moving oysters—A case study of global translocations of Ostrea edulis Linnaeus, 1758 (Mollusca: Bivalvia). *Ocean. Coast. Manag.* **2016**, *122*, 103–115. [CrossRef]

49. Ruesink, J.L.; Lenihan, H.S.; Trimble, A.C.; Heiman, K.W.; Micheli, F.; Byers, J.E.; Kay, M.C. Introduction of non-native oysters: Ecosystem effects and restoration implications. *Annu. Rev. Ecol. Syst.* **2005**, *36*, 643–689. [CrossRef]

50. Beaz-Hidalgo, R.; Romalde, S.; Figueras, M. Diversity and pathogenecity of Vibrio species in cultured bivalve molluscs. *Environ. Microbiol. Rep.* **2010**, *2*, 34–43. [CrossRef] [PubMed]

51. Azéma, P.; Lamy, J.-B.; Boudry, P.; Renault, T.; Travers, M.-A.; Dégremont, L. Genetic parameters of resistance to Vibrio aestuarianus, and OsHV-1 infections in the Pacific oyster, Crassostrea gigas, at three different life stages. *Genet. Sel. Evol.* **2017**, *49*, 23. [CrossRef] [PubMed]

Journal of
Marine Science and Engineering

Article

Detection of Infectious Hypodermal and Hematopoietic Necrosis Virus (IHHNV, Decapod Penstylhamaparvovirus 1) in Commodity Red Claw Crayfish (*Cherax quadricarinatus*) Imported into South Korea

Chorong Lee [1,†], Seong-Kyoon Choi [2,3,†], Hye Jin Jeon [1], Seung Ho Lee [1], Young Kyoon Kim [1], Song Park [2], Jin-Kyu Park [1], Se-Hyeon Han [4], Seulgi Bae [1], Ji Hyung Kim [5,*] and Jee Eun Han [1,*]

1   College of Veterinary Medicine, Kyungpook National University, Daegu 41566, Korea; crlee@jejunu.ac.kr (C.L.); jhj1125@cu.ac.kr (H.J.J.); tmdgh134@knu.ac.kr (S.H.L.); fighters30@knu.ac.kr (Y.K.K.); jinkyu820@knu.ac.kr (J.-K.P.); sgbae@knu.ac.kr (S.B.)
2   Core Protein Resources Center, DGIST, Daegu 42988, Korea; cskbest@dgist.ac.kr (S.-K.C.); cristaling@dgist.ac.kr (S.P.)
3   Division of Biotechnology, DGIST, Daegu 42988, Korea
4   Department of News-Team, Seoul Broadcasting System, Seoul 07574, Korea; vetman@sbs.co.kr
5   Infectious Disease Research Center, Korea Research Institute of Bioscience and Biotechnology, Daejeon 34141, Korea
*   Correspondence: kzh81@kribb.re.kr (J.H.K.); jehan@knu.ac.kr (J.E.H.); Tel.: +82-53-950-5972 (J.E.H.)
†   These authors contributed equally to this work.

Citation: Lee, C.; Choi, S.-K.; Jeon, H.J.; Lee, S.H.; Kim, Y.K.; Park, S.; Park, J.-K.; Han, S.-H.; Bae, S.; Kim, J.H.; et al. Detection of Infectious Hypodermal and Hematopoietic Necrosis Virus (IHHNV, Decapod Penstylhamaparvovirus 1) in Commodity Red Claw Crayfish (*Cherax quadricarinatus*) Imported into South Korea. *J. Mar. Sci. Eng.* **2021**, *9*, 856. https://doi.org/10.3390/jmse9080856

Academic Editor: Snježana Zrnčić

Received: 17 June 2021
Accepted: 5 August 2021
Published: 9 August 2021

Publisher's Note: MDPI stays neutral with regard to jurisdictional claims in published maps and institutional affiliations.

Abstract: Freshwater crayfish, which are cultivated in aquaculture, are economically important for food and ornamental purposes. However, relatively few studies have focused on potentially pathogenic viruses in crayfish compared to in penaeid shrimp. Commodity red claw crayfish (*Cherax quadricarinatus*; 400 crayfish in 10 batches) and red swamp crayfish (*Procambarus clarkii*; 40 crayfish in 2 batches) imported into South Korea from Indonesia and China were screened by PCR to detect infectious hypodermal and hematopoietic necrosis virus (IHHNV or Decapod penstylhamaparvovirus 1). IHHNV was detected in tissue samples pooled from nine out of ten batches of red claw crayfish imported from Indonesia. Phylogenetic analysis of PCR amplicons from representative pools clustered the IHHNV strain with infectious-type II sequences commonly detected in Southeast Asian countries rather than with type III strains detected previously in whiteleg shrimp (*Penaeus vannamei*) cultured in South Korea. IHHNV DNA was detected most frequently in the muscle (eight batches, 66.7% samples), followed by in the hepatopancreas (five batches, 41.7% samples) and gills tissue (three batches, 25.0% samples). These data suggest that red claw crayfish could be a potential carrier of the virus and that quarantine procedures must be strengthened in South Korea to avoid importing infectious types of IHHNV in commodity crustaceans such as red claw crayfish.

Keywords: infectious hypodermal and hematopoietic necrosis virus; infectious type; red claw crayfish; type II; reservoir

1. Introduction

The red claw crayfish *Cherax quadricarinatus* is a large, highly productive, and rapidly growing freshwater decapod crustacean that can live in diverse environments [1]. Since the mid-1980s, red claw crayfish aquaculture has grown rapidly in tropical and sub-tropical regions including South Africa, Zimbabwe, Japan, America, China, and Chile [2]. Interest in both aquaculture and aquarium trade has resulted in the species being translocated worldwide [3]. Although it is farmed mostly in extensive pond systems, intensive rearing systems are becoming common [2,4]. Pathogens in crayfish hosts have not been widely investigated, and such intensive production systems have increased the likelihood of disease impacts due to viruses such as white spot syndrome virus, *C. quadricarinatus* bacilliform virus, *Cherax giardiavirus*-like virus, spawner-isolated mortality virus, a putative gill parvovirus, reo-like virus, and *C. quadricarinatus* parvo-like virus [5–10].

Infectious hypodermal and hematopoietic necrosis virus (IHHNV, also named as *Decapod penstylhamaparvovirus* 1 for taxonomic consistency) is the smallest of the known crustacean viruses, with a non-enveloped icosahedral head and ssDNA genome of approximately 3.9 kb in length; the virus is a known shrimp pathogen causing cuticular deformities described as runt-deformity syndrome and had been listed as one of the notifiable crustacean pathogens by the World Organization for Animal Health (OIE) [11,12]. Until recently, a total of five genotypes (three infectious types including types I, II, and III and two non-infectious types including types A and B) of IHHNVs have been described [13]. Although the infectious type of IHHNV primarily infects penaeid shrimp [14], the virus can also infect the giant freshwater prawn *Macrobrachium rosenbergii*, grapsid crab *Hemigrapsus penicillatus*, and estuarine crab *Neohelice granulate* [14–16]. Natural and experimental infections of the red swamp crayfish *Procambarus clarkii* have also been reported [17–19], with an outbreak in wild *P. clarkii* at Weishan Lake, Shadong province in China resulting in nearly 100% mortality [17]. Additionally, disease transmission between shrimp and crayfish was confirmed by feeding IHHNV-infected *P. vannamei* tissue to *P. clarkii* [18]. Although some previous studies reported the presence of Cowdry type A inclusion bodies [20] and endogenous Brevidensovirus-like elements [21] in red claw crayfish, these studies did not provide clear evidence of IHHNV infection in the species.

In this study, we investigated the potential presence of IHHNV in commodity red claw crayfish and red swamp crayfish purchased from Korean fishery markets but originally imported from foreign countries (Indonesia and China) using conventional PCR and phylogenetic analysis.

2. Materials and Methods

2.1. Samples

Frozen *C. quadricarinatus* imported from Indonesia (400 crayfish in 10 batches, approximately 15–48 g) and *P. clarkii* imported from China (40 crayfish in 2 batches, approximately 52–63 g) were purchased from 9 retail fish markets in South Korea (Table 1). The samples were stored at −80 °C until processing.

Table 1. Crayfish species, country of origin, year collected, and PCR detection of IHHNV in DNA extracted from each of three different tissue types pooled from each batch of crayfish.

Sample	Species	Origin	Year/Month	IHHNV Detection		
				Muscle	HP *	Gill
20-002	*Cherax quadricarinatus*	Indonesia	2019/03	+ **	+ **	-
20-003	*Procambarus clarkii*	China	2018/10	-	-	-
20-004	*Cherax quadricarinatus*	Indonesia	2019/01	+ **	+ **	+ **
20-005	*Cherax quadricarinatus*	Indonesia	2019/04	+ ***	-	-
20-006	*Cherax quadricarinatus*	Indonesia	2019/06	-	-	-
20-007	*Cherax quadricarinatus*	Indonesia	2019/02	+ **	-	+ **
20-008	*Cherax quadricarinatus*	Indonesia	2020/01	+ **	+ **	+ **
20-009	*Cherax quadricarinatus*	Indonesia	2019/04	+ **	+ **	-
20-010	*Cherax quadricarinatus*	Indonesia	2019/04	+ **	-	-
20-011	*Cherax quadricarinatus*	Indonesia	2019/02	+ ***	-	-
20-012	*Procambarus clarkii*	China	2018/10	-	-	-
20-013	*Cherax quadricarinatus*	Indonesia	2018/12	-	+ ***	-

* HP, hepatopancreas. ** Sequenced samples. *** Not sequenced due to low level of amplification.

2.2. DNA Extraction

Five crayfish were randomly selected from each batch, and their muscles (~6 mg from each crayfish for a total of 30 mg per batch), hepatopancreases (~6 mg from each crayfish for a total of 30 mg per batch), and gills (~6 mg from each crayfish for a total of 30 mg per batch) were collected. DNA was extracted from different tissues with DNeasy Blood & Tissue kits (69506, Qiagen, Hilden, Germany) according to the manufacturer's instructions. Each DNA sample was processed separately, after which the five samples for each organ were pooled, resulting in three samples (one for each organ type) for PCR analysis.

2.3. IHHNV PCR

According to the recommendation of the OIE [22], the viral DNA was amplified by IHHNV-389F/389R PCR primers which targeting the nonstructural protein-coding region of the IHHNV genome [23]. The reaction samples (25 μL) contained AccuPower® PCR PreMix (K-2016, Bioneer, Daejeon, South Korea), 1 μL of each DNA extract, and 1 μL (10 pmole) of each primer. The thermal cycling conditions were as follows: 95 °C for 5 min followed by 35 cycles of 95 °C for 30 s, 60 °C for 30 s, and 72 °C for 30 s, followed by 72 °C for 5 min. The available PCR-positive amplicons were sequenced at Bioneer Inc. (Daejeon, Korea).

2.4. Sequence Comparison & Phylogenetic Analysis

The obtained nucleotide and its deduced amino acid sequences of presumptive IHHNV DNAs from the crayfish samples were compared with other available IHHNV sequences in the GenBank database using BLAST searches (available online: www.ncbi.nlm.nih.gov/BLAST (accessed on 1 March 2021)), which confirmed that the sequences resided within the IHHNV non-structural protein-coding region. The two IHHNV sequences from the hepatopancreas batches of *C. quadricarinatus* were selected and used for further phylogenetic analysis for these reasons; first, sample 20-002 showed multiple infections with white spot syndrome virus [24] and was selected for further analysis. Second, all the tested tissues (muscle, hepatopancreas, and gills) of sample 20-004 were positive for IHHNV detection and preferentially selected for further analysis. Trimmed sequences of IHHNV strains from different geographical origins and hosts were aligned using ClustalX (ver. 2.1) [25] and BioEdit Sequence Alignment Editor (ver. 7.1.0.3) [26]. A maximum-likelihood phylogenetic tree was constructed using the Jukes–Cantor model and 1000 bootstrap replicates in MEGA-X ver. 10.0 [27].

2.5. GenBank Accessions

The obtained partial sequences of the nonstructural protein-coding region in IHHNV detected from *C. quadricarinatus* hepatopancreas DNA of sample batches 20-002 and 20-004 were deposited in the GenBank database under accession numbers MT543324 and MT543323, respectively.

3. Results and Discussion

Although recent studies showed that *P. clarkii* can be infected by IHHNV both naturally and experimentally [17–19], the susceptibility of *C. quadricarinatus* to the viral infection remains unclear. Here, we detected IHHNV DNA in nine out of ten sample batches of *C. quadricarinatus* collected from Korean fish markets by PCR using 389F/389R primers (Table 1). Although IHHNV was not detected in the tissues of the two batches of *P. clarkii* imported from China, the sample number was small and may not be representative of the IHHNV infection status of this species farmed across different regions in China.

Among the different tissue types pooled from frozen *C. quadricarinatus* collected from fish markets, PCR detected IHHNV DNA in eight out of ten batches of muscle tissue, five out of ten batches of hepatopancreas tissue, and three out of ten batches of gill tissue. Vial DNA was not detected in one out of ten batches in any of the three tissue types (Table 1). The sequenced PCR-positive amplicons of presumptive IHHNV DNAs showed >99% sequence identity within themselves. As a result of PCR analysis, a representative sample positively detected for IHHNV DNA in muscle tissue is shown in Figure 1. In penaeid shrimp, IHHNV primarily targets tissues of ectodermal and mesodermal origin, with the principal target organs including the gills, cuticular epithelium (or hypodermis), all connective tissues, hematopoietic tissues, the lymphoid organ, antennal gland, and the ventral nerve cord, its branches, and its ganglia, but not enteric tissues such as the hepatopancreas, midgut, or its caeca [12,28,29]. IHHNV was detected in the hepatopancreas of *C. quadricarinatus* as described previously for IHHNV infection in *P. clarkii*, possibly because the virus replicated in either internal connective tissue cells or hemocytes circulating in the hemolymph, or because virus particles circulated freely in the hemolymph [18,19].

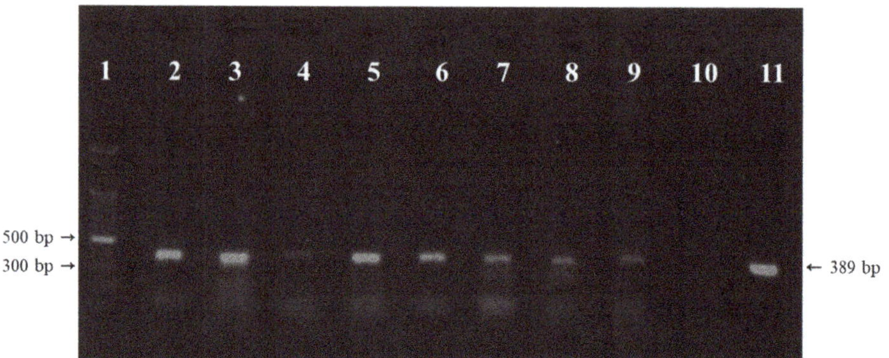

Figure 1. Agarose gel showing the 389-bp DNA amplified by IHHNV 389F/389R PCR primers using DNA extracts of muscle tissue pooled from imported red claw crayfish. Lane 1, 100-bp DNA maker (D-1030, Bioneer, Daejeon, South Korea); Lane 2, sample 20-002; Lane 3, sample 20-004; Lane 4, sample 20-005; Lane 5, sample 20-007; Lane 6, 20-008; Lane 7, sample 20-009; Lane 8, sample 20-010; Lane 9, sample 20-011; Lane 10, negative control; Lane 11, positive control.

To date, two distinct types of infectious IHHNV emerging in global shrimp culture farms have been identified as type II (primarily from Southeast Asia) and type III (primarily from the Americas and East Asia) [30,31]. Non-infectious types of endogenous IHHNV-related sequences have been found integrated in the genome of *P. monodon* from Africa and Australia [23], and a similar phenomenon was reported in *C. quadricarinatus* from Australia. Brevidensovirus-like elements showing high DNA homology but relatively low identities in its deduced amino acids with IHHNV were inserted into the crayfish genomes; however, their exact function remains unclear [21]. Moreover, the 389F/389R primer set used can amplify non-infectious endogenous types of IHHNV [32]. Therefore, we conducted further sequence comparisons and phylogenetic analyses using IHHNV DNA obtained from *C. quadricarinatus* to discriminate infectious and non-infectious endogenous types and also determine the type of the infectious virus.

The IHHNV sequences obtained in this study exhibited high similarities (>96% and >98% identities each) with non-structural protein 1 and non-structural protein 2 of the reference IHHNV strain (Decapod penstylhamaparvovirus 1, NC_039043.1) detected in *P. stylirostris* from the Gulf of California in 1998 [33], as well as with other IHHNV strains detected in shrimp and mollusks from diverse geographical locations. Moreover, in contrast to a previous study [21], the deduced amino acids of the PCR amplicons obtained showed high similarity (>96% and >98% identities, each) with non-structural protein 1 and non-structural protein 2 of the reference IHHNV strain (NC_039043.1). Phylogenetic analyses of representative 389F/389R PCR amplicons (~380 bp) from 2 batches of *C. quadricarinatus* (samples 20-002 (MT543324) and 20-004 (MT543323)) from Indonesia clustered closely together with type II strains mainly reported from Southeast Asian countries (Figure 2). Considering the evidence that (i) not all the *C. quadricarinatus* samples tested in this study were positive in the PCR analysis, (ii) the nucleotide sequences of the positive PCR amplicons exhibited high similarities (>96%) to the nonstructural protein-coding region in IHHNV, and (iii) the sequences were phylogenetically clustered with the IHHNV strain with infectious-type II, the IHHNV DNA obtained from *C. quadricarinatus* in the current study was the infectious type rather than non-infectious endogenous IHHNV-related sequences. As only IHHNV type III strains were previously detected in South Korea [29,30], the IHHNV-positive imported commodity *C. quadricarinatus* is a potential reservoir that can inadvertently release this type into wild and farmed shrimp and crayfish in South Korea.

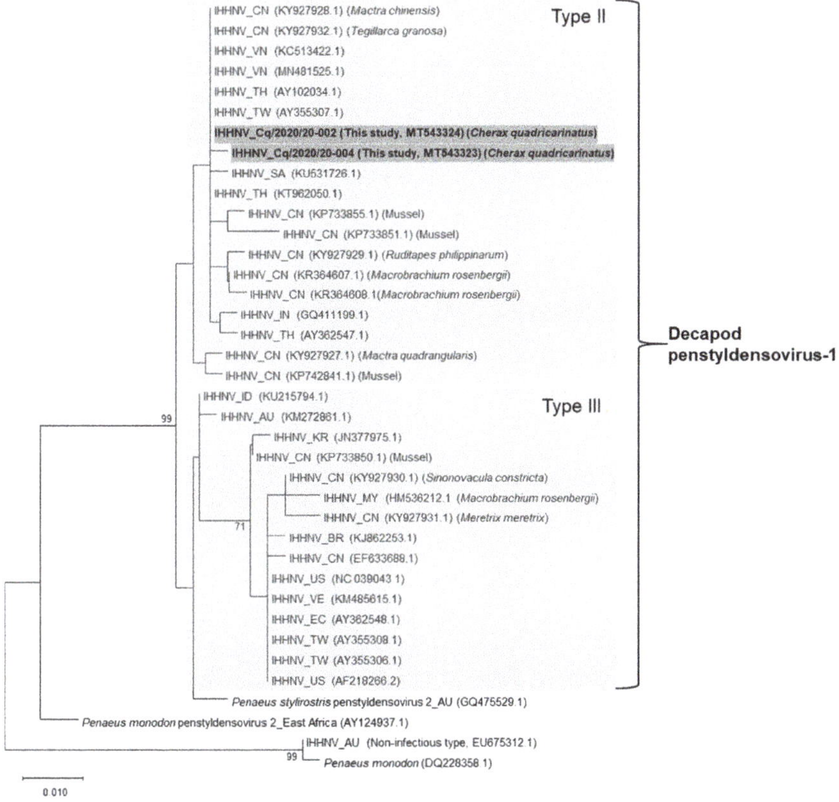

Figure 2. Phylogenetic tree based on the nucleotide sequences of the partial nonstructural protein-coding region (~380 bp) in the representative IHHNV strains including the strain detected in two batches of Cherax quadricarinatus (Sample 20-002; MT543324; Sample 20-004, MT543323) imported from Indonesia. The maximum-likelihood tree also included non-infectious endogenous IHHNV types (EU675312.1 and DQ228358.1), which were used as an outgroup. Bootstrap values were calculated from 1000 replicates. The scale bar (0.01) represents 1 substitution per 100 nucleotides.

The detection of IHHNV in the imported commodity red claw crayfish indicates that this virus poses a threat to crustacean aquaculture industries in South Korea [34,35], particularly as viruses can remain viable in frozen shrimp tissue and be transmitted via feeding of infected tissue to other crustaceans [36]. Stricter controls and monitoring of imported crayfish for crustacean viruses are warranted. In addition, an experimental challenge of *C. quadricarinatus* with IHHNV is necessary to determine its histopathological and mortality characteristics.

4. Conclusions

Recently, crayfish have become economically important crustaceans in the aquaculture industry for food and ornamental purposes. However, transboundary movements can cause significant environmental disturbance and provide opportunities for inadvertent translocation of disease-causing pathogens to new locations. Although infection by IHHNV was recently confirmed in the red swamp crayfish *P. clarkii*, it remains unknown whether red claw crayfish (*C. quadricarinatus*) are susceptible to IHHNV infection. Here, we detected IHHNV infection in *C. quadricarinatus* by identifying the viral DNA in tissue samples of the commodity imported into South Korea as the IHHNV type II strain. These results indicate that the red claw crayfish could be a potential carrier of the infectious IHHNV and suggest the need for stricter monitoring of imported crayfish crustacean viruses that may impact cultured and wild populations of crustacean in South Korea.

Author Contributions: Conceptualization, C.L. and J.E.H.; data curation, S.-K.C., H.J.J. and S.P.; formal analysis, H.J.J., S.H.L. and Y.K.K.; funding acquisition, J.E.H. and J.H.K.; investigation, C.L., J.-K.P., S.-H.H., S.B. and J.H.K.; methodology, C.L., J.E.H. and J.H.K.; project administration, J.E.H.; resources, J.E.H.; supervision, J.E.H. and J.H.K.; writing—original draft, C.L., S.-K.C., J.E.H. and J.H.K.; and writing—review and editing, J.E.H. and J.H.K. All authors have read and agreed to the published version of the manuscript.

Funding: This work was supported by the National Research Foundation (NRF) (NRF-2019R1C1C1006212 and 2020R1I1A2068827), KRIBB Research Initiative Program, and DGIST R&D Program of the Ministry of Science and ICT (2020010096).

Institutional Review Board Statement: Not applicable.

Informed Consent Statement: Not applicable.

Data Availability Statement: The obtained partial sequences of the nonstructural protein-coding region in IHHNV detected from *C. quadricarinatus* hepatopancreas DNAs of sample batches 20-002 and 20-004 were deposited in the GenBank database under accession numbers MT543324 and MT543323, respectively.

Conflicts of Interest: The authors declare no conflict of interest.

References

1. Yang, H.; Wei, X.; Wang, R.; Zeng, L.; Yang, Y.; Huang, G.; Shafique, L.; Ma, H.; Lv, M.; Ruan, Z.; et al. Transcriptomics of *Cherax quadricarinatus* hepatopancreas during infection with Decapod iridescent virus 1 (DIV1). *Fish Shellfish Immunol.* **2020**, *98*, 832–842. [CrossRef] [PubMed]
2. Saoud, I.P.; Ghanawi, J.; Thompson, K.R.; Webster, C.D. A review of the culture and diseases of redclaw crayfish *Cherax quadricarinatus* (von Martens 1868). *J. World Aquac. Soc.* **2013**, *44*, 1–29. [CrossRef]
3. Liu, S.; Qi, C.; Jia, Y.; Gu, Z.; Li, E. Growth and intestinal health of the red claw crayfish, *Cherax quadricarinatus*, reared under different salinities. *Aquaculture* **2020**, *524*, 735256. [CrossRef]
4. Edgerton, B.F.; Evans, L.H.; Stephens, F.J.; Overstreet, R.M. Synopsis of freshwater crayfish diseases and commensal organisms. *Aquaculture* **2002**, *206*, 57–135. [CrossRef]
5. Dragičević, P.; Bielen, A.; Petrić, I.; Hudina, S. Microbial pathogens of freshwater crayfish: A critical review and systematization of the existing data with directions for future research. *J. Fish Dis.* **2021**, *44*, 221–247. [CrossRef]
6. Anderson, I.G.; Prior, H.C. Baculovirus infections in the mud crab, *Scylla serrata*, and a freshwater crayfish, *Cherax quadricarinatus*, from Australia. *J. Invertebr. Pathol.* **1992**, *60*, 265–273. [CrossRef]
7. Edgerton, B.; Owens, L.; Giasson, B.; De Beer, S. Description of a small dsRNA virus from freshwater crayfish *Cherax quadricarinatus*. *Dis. Aquat. Org.* **1994**, *18*, 63–69. [CrossRef]
8. Owens, L.; McElnea, C. Natural infection of the redclaw crayfish *Cherax quadricarinatus* with presumptive spawner-isolated mortality virus. *Dis. Aquat. Org.* **2000**, *40*, 219–223. [CrossRef] [PubMed]
9. Edgerton, B.F.; Webb, R.; Anderson, I.G.; Kulpa, E.C. Description of a presumptive hepatopancreatic reovirus, and a putative gill parvovirus, in the freshwater crayfish *Cherax quadricarinatus*. *Dis. Aquat. Org.* **2000**, *41*, 83–90. [CrossRef] [PubMed]
10. Bowater, R.O.; Wingfield, M.; Fisk, A.; Condon, K.M.; Reid, A.; Prior, H.; Kulpa, E.C. A parvo-like virus in cultured redclaw crayfish *Cherax quadricarinatus* from Queensland, Australia. *Dis. Aquat. Org.* **2002**, *50*, 79–86. [CrossRef]
11. Yu, J.Y.; Yang, N.; Hou, Z.H.; Wang, J.J.; Li, T.; Chang, L.R.; Fang, Y.; Yan, D.C. Research progress on hosts and carriers, prevalence, virulence of infectious hypodermal and hematopoietic necrosis virus (IHHNV). *J. Invertebr. Pathol.* **2021**, *183*, 107556. [CrossRef] [PubMed]

12. Lightner, D.V.; Redman, R.M. Shrimp diseases and current diagnostic methods. *Aquaculture* **1998**, *164*, 201–220. [CrossRef]

13. Shen, H.; Zhang, W.; Shao, S. Phylogenetic and recombination analysis of genomic sequences of IHHNV. *J. Basic Microbiol.* **2015**, *55*, 1048–1052. [CrossRef] [PubMed]

14. Yang, B.; Song, X.L.; Huang, J.; Shi, C.Y.; Liu, L. Evidence of existence of infectious hypodermal and hematopoietic necrosis virus in penaeid shrimp cultured in China. *Vet. Microbiol.* **2007**, *120*, 63–70. [CrossRef] [PubMed]

15. Nita, M.H.; Kua, B.C.; Bhassu, S.; Othman, R.Y. Detection and genetic profiling of infectious hypodermal and haematopoietic necrosis virus (IHHNV) infections in wild berried freshwater prawn, *Macrobrachim rosenbergii* collected for hatchery production. *Mol. Biol. Rep.* **2012**, *39*, 3785–3790. [CrossRef] [PubMed]

16. Cavalli, L.S.; Batista, C.R.; Nornberg, B.F.; Mayer, F.Q.; Seixas, F.K.; Romano, L.A.; Marins, L.F.; Abreu, P.C. Natural occurrence of white spot syndrome virus and infectious hypodermal and hematopoietic necrosis virus in *Neohelice granulata* crab. *J. Invertebr. Pathol.* **2013**, *114*, 86–88. [CrossRef]

17. Chen, B.K.; Dong, Z.; Liu, D.P.; Yan, Y.B.; Pang, N.Y.; Nian, Y.Y.; Yan, D.C. Infectious hypodermal and haematopoietic necrosis virus (IHHNV) infection in freshwater crayfish *Procambarus clarkii*. *Aquaculture* **2017**, *477*, 76–79. [CrossRef]

18. Chen, B.-K.; Dong, Z.; Pang, N.-Y.; Nian, Y.-Y.; Yan, D.-C. A novel real-time PCR approach for detection of infectious hypodermal and haematopoietic necrosis virus (IHHNV) in the freshwater crayfish *Procambarus clarkii*. *J. Invertebr. Pathol.* **2018**, *157*, 100–103. [CrossRef]

19. Nian, Y.-Y.; Chen, B.-K.; Wang, J.-J.; Zhong, W.-T.; Fang, Y.; Li, Z.; Zhang, Q.-S.; Yan, D.-C. Transcriptome analysis of *Procambarus clarkii* infected with infectious hypodermal and haematopoietic necrosis virus. *Fish Shellfish Immunol.* **2020**, *98*, 766–772. [CrossRef]

20. Romero, X.; Jiménez, R. Histopathological survey of diseases and pathogens present in redclaw crayfish, *Cherax quadricarinatus* (Von Martens), cultured in Ecuador. *J. Fish Dis.* **2002**, *25*, 653–667. [CrossRef]

21. La Fauce, K.A.; Elliman, J.; Bowater, R.O.; Owens, L.; Rusaini. Endogenous Brevidensovirus-like elements in *Cherax quadricarinatus*: Friend or foe? *Aquaculture* **2013**, *396*, 136–145.

22. OIE (World Organisation for Animal Health). Manual of Diagnostic Tests for Aquatic Animals. 2021. Available online: https://www.oie.int/en/what-we-do/standards/codes-and-manuals/aquatic-manual-online-access/ (accessed on 1 March 2021).

23. Tang, K.F.; Lightner, D.V. Infectious hypodermal and hematopoietic necrosis virus (IHHNV)-related sequences in the genome of the black tiger prawn *Penaeus monodon* from Africa and Australia. *Virus Res.* **2006**, *118*, 185–191. [CrossRef] [PubMed]

24. Lee, C.; Kim, J.H.; Choi, S.K.; Jeon, H.J.; Lee, S.H.; Kim, B.G.; Kim, Y.K.; Lee, K.J.; Han, J.E. Detection of infectious white spot syndrome virus in red claw crayfish (*Cherax quadricarinatus*) and red swamp crayfish (*Procambarus clarkii*) imported into Korea. *Aquaculture* **2021**, *544*, 737117. [CrossRef]

25. Larkin, M.A.; Blackshields, G.; Brown, N.P.; Chenna, R.; McGettigan, P.A.; McWilliam, H.; Valentin, F.; Wallace, I.M.; Wilm, A.; Lopez, R.; et al. Clustal W and Clustal X version 2.0. *Bioinformatics* **2007**, *23*, 2947–2948. [CrossRef] [PubMed]

26. Hall, T.A. BioEdit: A user-friendly biological sequence alignment editor and analysis program for Windows 95/98/NT. *Nucleic Acids Symp. Ser.* **1999**, *41*, 95–98.

27. Kumar, S.; Stecher, G.; Li, M.; Knyaz, C.; Tamura, K. MEGA X: Molecular evolutionary genetics analysis across computing platforms. *Mol. Biol. Evol.* **2018**, *35*, 1547–1549. [CrossRef]

28. Lightner, D.V. *A Handbook of Shrimp Pathology and Diagnostic Procedures for Diseases of Cultured Penaeid Shrimp*; World Aquaculture Society: Baton Rouge, LA, USA, 1996.

29. Lightner, D.V. Status of shrimp diseases and advances in shrimp health management. In *Diseases in Asian Aquaculture VII*; Fish Health Section; Asian Fisheries Society: Selangor, Malaysia, 2011; pp. 121–134.

30. Kim, J.H.; Kim, H.K.; Nguyen, V.G.; Park, B.K.; Choresca, C.H.; Shin, S.P.; Han, J.E.; Jun, J.W.; Park, S.C. Genomic sequence of infectious hypodermal and hematopoietic necrosis virus (IHHNV) KLV-2010-01 originating from the first Korean outbreak in cultured *Litopenaeus vannamei*. *Arch. Virol.* **2012**, *157*, 369–373. [CrossRef] [PubMed]

31. Kim, J.H.; Choresca, C.H., Jr.; Shin, S.P.; Han, J.E.; Jun, J.W.; Han, S.Y.; Park, S.C. Detection of infectious hypodermal and hematopoietic necrosis virus (IHHNV) in *Litopenaeus vannamei* shrimp cultured in South Korea. *Aquaculture* **2011**, *313*, 161–164. [CrossRef]

32. Tang, K.F.; Navarro, S.A.; Lightner, D.V. PCR assay for discriminating between infectious hypodermal and hematopoietic necrosis virus (IHHNV) and virus-related sequences in the genome of *Penaeus monodon*. *Dis. Aquat. Org.* **2007**, *74*, 165–170. [CrossRef]

33. Shike, H.; Dhar, A.K.; Burns, J.C.; Shimizu, C.; Jousset, F.X.; Klimpel, K.R.; Bergoin, M. Infectious hypodermal and hematopoietic necrosis virus of shrimp is related to mosquito brevidensoviruses. *Virology* **2000**, *277*, 167–177. [CrossRef]

34. Lightner, D.V.; Redman, R.M.; Poulos, B.T.; Nunan, L.M.; Mari, J.L.; Hasson, K.W. Risk of spread of penaeid shrimp viruses in the Americas by the international movement of live and frozen shrimp. *Rev. Sci. Tech. OIE* **1997**, *16*, 146–160. [CrossRef]

35. Park, S.C.; Choi, S.K.; Han, S.H.; Park, S.; Jeon, H.J.; Lee, S.C.; Kim, K.Y.; Lee, Y.S.; Kim, J.H.; Han, J.E. Detection of infectious hypodermal and hematopoietic necrosis virus and white spot syndrome virus in whiteleg shrimp (*Penaeus vannamei*) imported from Vietnam to South Korea. *J. Vet. Sci* **2020**, *21*, e31. [CrossRef] [PubMed]

36. Jones, B. Transboundary movement of shrimp viruses in crustaceans and their products: A special risk? *J. Invertebr. Pathol.* **2012**, *110*, 196–200. [CrossRef] [PubMed]

Journal of
*Marine Science
and Engineering*

Article

The Isolation of *Vibrio crassostreae* and *V. cyclitrophicus* in Lesser-Spotted Dogfish (*Scyliorhinus canicula*) Juveniles Reared in a Public Aquarium

Mattia Tomasoni [1,†], Giuseppe Esposito [1,*,†], Davide Mugetti [1,*], Paolo Pastorino [1], Nadia Stoppani [1], Vasco Menconi [1], Flavio Gagliardi [2], Ilaria Corrias [2], Angela Pira [2], Pier Luigi Acutis [1], Alessandro Dondo [1], Marino Prearo [1] and Silvia Colussi [1]

[1]   Istituto Zooprofilattico Sperimentale del Piemonte, Liguria e Valle d'Aosta, Via Bologna 148, 10154 Torino, Italy; mattia.tomasoni@izsto.it (M.T.); paolo.pastorino@izsto.it (P.P.); nadia.stoppani@izsto.it (N.S.); vasco.menconi@izsto.it (V.M.); pierluigi.acutis@izsto.it (P.L.A.); alessandro.dondo@izsto.it (A.D.); marino.prearo@izsto.it (M.P.); silvia.colussi@izsto.it (S.C.)

[2]   Acquario di Cala Gonone, Via La Favorita, 08022 Dorgali, Italy; flaviogagliardi@panaque.com (F.G.); acquariologia2@acquariocalagonone.it (I.C.); acquariologia@acquariocalagonone.it (A.P.)

*   Correspondence: giuseppe.esposito@izsto.it (G.E.); davide.mugetti@izsto.it (D.M.); Tel.: +39-011-268-6251 (G.E.); +39-011-268-6367 (D.M.)

†   These authors contributed equally to this work.

Abstract: The genus *Vibrio* currently contains 147 recognized species widely distributed, including pathogens for aquatic organisms. *Vibrio* infections in elasmobranchs are poorly reported, often with identifications as *Vibrio* sp. and without detailed diagnostic insights. The purpose of this paper is the description of the isolation and identification process of *Vibrio* spp. following a mortality event of *Scyliorhinus canicula* juvenile reared in an Italian public aquarium. Following investigations aimed at excluding the presence of different pathogens of marine fish species (parasites, bacteria, Betanodavirus), several colonies were isolated and subjected to species identification using the available diagnostic techniques (a biochemical test, MALDI-TOF MS, and biomolecular analysis). Discrepancies were observed among the methods; the limits of biochemistry as a unique tool for *Vibrio* species determination were detected through statistical analysis. The use of the *rpoB* gene, as a diagnostic tool, allowed the identification of the isolates as *V. crassostreae* and *V. cyclotrophicus*. Although the pathogenic role of these microorganisms in lesser-spotted dogfish juveniles has not been demonstrated, and the presence of further pathogens cannot be excluded, this study allowed the isolation of two *Vibrio* species in less-studied aquatic organisms, highlighting the weaknesses and strengths of the different diagnostic methods applied.

Keywords: biochemical characterization; captive sharks; diagnostic techniques; opportunistic pathogens; Vibrionaceae

Citation: Tomasoni, M.; Esposito, G.; Mugetti, D.; Pastorino, P.; Stoppani, N.; Menconi, V.; Gagliardi, F.; Corrias, I.; Pira, A.; Acutis, P.L.; et al. The Isolation of *Vibrio crassostreae* and *V. cyclitrophicus* in Lesser-Spotted Dogfish (*Scyliorhinus canicula*) Juveniles Reared in a Public Aquarium. *J. Mar. Sci. Eng.* **2022**, *10*, 114. https://doi.org/10.3390/jmse10010114

Academic Editor: Snježana Zrnčić

Received: 10 November 2021
Accepted: 13 January 2022
Published: 15 January 2022

Publisher's Note: MDPI stays neutral with regard to jurisdictional claims in published maps and institutional affiliations.

1. Introduction

The bacteria of the Vibrionaceae family are Gram-negative curved rods that globally occur in marine, estuarine and freshwater ecosystems. They occupy habitats ranging from the deep-sea to shallow aquatic environments [1]. Among them, some species are important for natural systems, including carbon cycle and osmoregulation, and as free-living inhabitants in the water column or associated with particulate matter [2]. The genus *Vibrio* currently contains 147 recognized species widely distributed in aquatic environments [3]. Some vibrios cause water- and seafood-related outbreaks of gastrointestinal infections in humans [4], and many can be pathogenic for marine vertebrates [5,6] and invertebrates [7,8]. Paillard et al. recognized that the emergence of vibrios as etiological agents of diseases is likely to increase over the coming years due to ocean warming [9]. Classical vibriosis is generally characterized by lethargic movement of affected fish [10–12], the presence of various skin ulcerations [12–15], rotting fins [14,16,17], and pigmentation of the body [10,13,14,18]. Internal organs appear enlarged, hemorrhagic and congested, particularly the liver, kidney, and spleen [11,12,18]. Clinical signs demonstrated during vibriosis outbreaks are different from case to case and are influenced by multiple factors such as host species, age, exposure time, and strain virulence factors. Epidemiologically, vibriosis is often acute in young fish larvae and fry. The disease spread rapidly following infection and most infected young fish die without showing clinical signs. Thus, sudden death is often noticed in young fish that died with less severe or no external symptoms. In contrast, infected adult fish usually develop a chronic infection with evident skin ulcerations and pigmentation [19]. Stress is a major factor in increasing fish susceptibility to pathogens [15]. Regarding marine fish aquaculture, the most hazardous species of the Vibrionaceae family are *V. anguillarum*, *V. ordalii*, *V. salmonicida*, *V. vulnificus*, *V. alginolyticus*, *V. harveyi*, *V. parahaemolyticus*, *V. ponticus* for the genera *Vibrio*, and *Photobacterium damselae* subsp. *damselae* (formerly known as *Vibrio damselae*) [20–24]. Lesser-reported species of the genus *Vibrio* in other marine organisms are poorly described; this is presumably due to the high diversity of Vibrionaceae in environments and from the natural environment itself, which does not present the typical stressors of intensive farming. Grimes et al. [25] reported a mortality case of a single brown shark (*Carcharhinus plumbeus*) reared in captivity due to an infection by a *Vibrio* species, identified as *V. carchariae*. However, this isolate was later recognized as a *V. harveyi* strain based on whole-genome sequencing data [26].

Elasmobranchs are considered a fundamental element in the trophic chain of marine environments [27,28]. Currently, the population of elasmobranchs in the Mediterranean Sea is in decline due to habitat degradation and consequent to the direct impacts of fishing [29–31]. At the time of writing, 41% of the species are considered at risk (classified as either critically endangered, endangered, or vulnerable) by the regional assessment of the International Union for the Conservation of Nature (IUCN), and 33% are considered as lacking in data [32]. Focusing on the lesser-spotted dogfish *Scyliorhinus canicula* (Linnaeus, 1758), a small demersal shark (Carcharhiniformes: Scyliorhinidae) classified as LC by IUCN, it is a broad generalist, and in terms of diet and habitat requirements, it is an opportunistic feeder, and cannibalism is frequently recorded [33,34]. It is also possible to find this species in public aquaria because of its easy reproduction in captivity [35]. However, artificial housing conditions are often linked to the onset of diseases, and limited literature is available for sharks. To date, there are few reports on infectious diseases affecting sharks reported in the literature [36,37]. For this reason, this study aimed to clarify the cause of tank mortality that occurred in *S. canicula* juveniles reared in an Italian public aquarium.

2. Materials and Methods

2.1. Outbreak Description and Conditions of Sampling Procedure

Once laid, lesser-spotted dogfish eggs were moved from the exhibition tank to interconnecting tanks in a recirculated aquaculture system (RAS) (1 m^3 each), in accordance with the current legislation on animal welfare. The water introduced into a closed circuit, technically known as RAS, was subjected to different types of treatment—namely, mechanical, biological, thermal, gaseous, chemical rebalancing, and knocking down the bacterial charge. Specifically, disinfection of water occurred by means of special fluorescent lamps capable of producing ultraviolet (UV) light with wavelengths included in the UV–C band (λ 100–280 nm) [38]. Since organisms are exposed to UV light, UV rays affect nucleic acids (DNA and RNA), damaging the aromatic structures with the formation of double bonds in stable pyrimidine molecules and preventing the cellular development of microbes [39]. Only microorganisms such as algae, viruses, bacteria, and other pathogens floating in the water are impacted. Nevertheless, nitrifying bacteria (with key roles in the nitrogen cycle) are not carried to the UV sterilizer and therefore do not come in range and will not be lost. The following water physicochemical parameters were measured by a multiparametric field probe (HACH, HQd Field Case): temperature (17 ± 1 °C); salinity (36 ± 1 PSU); dissolved oxygen (8.5 ± 0.4 mg/L); pH (8.0 ± 0.3). The specimens (Figure 1) showed the first symptoms a few months after hatching, i.e., anomalies in swimming, fast breathing, and lack of appetite. For the analysis, symptomatic and moribund fish were sent inside a double plastic bag (containing $^1/_3$ water and 2/3 air/oxygen) and sent refrigerated to the fish diseases laboratory of the Istituto Zooprofilattico Sperimentale of Piemonte, Liguria and Valle d'Aosta, Turin, Italy.

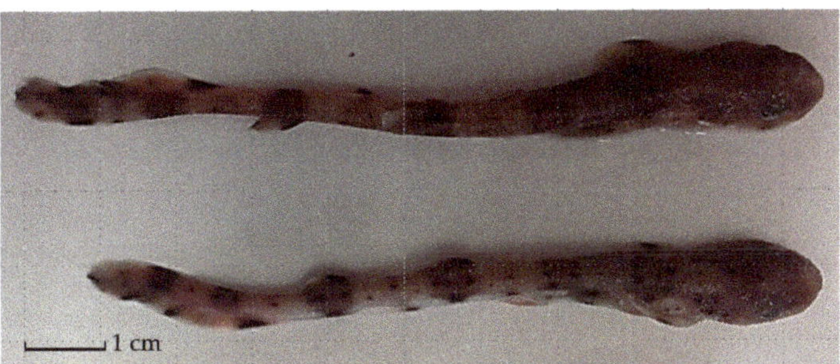

Figure 1. Specimens of lesser-spotted dogfish (*Scyliorhinus canicula*) analyzed during this survey.

2.2. Necropsy

In the laboratory, the fish were euthanatized using an overdose of ethyl 3-aminobenzoate methanesulfonate (MS-222, Sigma Aldrich, St. Louis, USA) following the current legislation. Before necropsy, biometrical parameters of each animal were recorded (total length (TL) and total weight (TW)). The fish were then macroscopically examined to highlight external alterations. After dissection with sterile tools, the coelomic cavity and the internal organs were subjected to visual inspection for the evaluation of anatomopathological changes.

2.3. Parasitological Examination

A general approach was used for parasitological examination in order to detect eso- and ectoparasites. Gills, skin, visceral cavity, and digestive tract were analyzed, both macroscopically and microscopically, with the use of a microscope. The microscope used for the analysis is an OLYMPUS BX40, with magnification ranging from 4× to 40×.

2.4. Virological Examination

For virological analysis, a sample of a portion of the brain was taken for the search for viral encephalopathy and retinopathy virus (Betanodavirus). The total RNA was extracted using the AllPrep DNA/RNA Micro Kit (QIAGEN, Hilden, Germany) following the manufacturer's protocol. Subsequently, cDNA was synthesized from extracted RNA using the QuantiTect Reverse Transcription Kit (QIAGEN, Hilden, Germany). A real-time RT–PCR targeting the viral encephalopathy and retinopathy virus RNA2 portion was used for viral diagnosis, following the protocol described by Panzarin et al. [40]. The reactions were carried out using the BioRad CFX96 Touch Real-Time PCR Detection System (Bio-Rad, Hercules, USA), and the data were analyzed by Bio-Rad CFX Maestro Software version 1.1.

2.5. Bacteriological and Biochemical Analyses

Bacteriological sampling was carried out from the head kidney, brain, and blood. The organs samples were taken using a 1 µL sterile calibrated loop, while blood sampling was performed immediately after the euthanasia procedure, taking it directly from the sharks' hearts with an insulin syringe, inoculated on plates, and spread using a 1 µL sterile loop. The first isolation was performed on Columbia blood agar (BA), tryptic soy agar (TSA) supplemented with 2% NaCl, and Monsur medium (thiosulfate citrate bile sucrose, TCBS); the latter medium was used for its selectivity toward Vibrionaceae. Plates were incubated at 22 ± 2 °C for a total of 72 h, and growth was checked daily. After growth, colonies were cloned on BA for identification analyses. After Gram staining, pure isolates were identified by API® 20E tests, matrix-assisted laser desorption ionization–time of flight mass spectrometry (MALDI–TOF MS) using VITEK-MS (BioMerieux, Marcy-l'Étoile, France), and molecular analyses. API® tests were also used for the biochemical characterization of strains. API® galleries were incubated for 48 h at 22 ± 2 °C; tests were read every 24 h, and reagents were added according to the manufacturer's instructions. The results were issued using the specific API® test reading software (APIWEB™) available on the BioMerieux website (https://apiweb.biomerieux.com, accessed on 30 September 2021).

2.6. Biomolecular Analysis for Bacterial Identification

DNA extraction was performed using the boiling and freeze–thawing protocol as described by Pastorino et al. [41]. RNA polymerase beta subunit (*rpoB*) gene was amplified to discriminate among different *Vibrio* species. Specific primers for the *rpoB* gene, developed by Ki et al. [42], and the protocol suggested by the same authors were used. Amplicons were run on 2% GelGreen (Biotium, Landing Pkwy, USA) stained agarose gel and then visualized under UV exposure. A 50–2000 kb ladder (Amplisize Molecular Ruler, Bio-Rad, Hercules, USA) was used as a molecular marker. The amplicons were purified with an ExtractMe DNA Clean-Up and Gel-Out Kit (Blirt, Gdańsk, Poland), according to the manufacturer's instructions. The purified PCR products were bidirectionally sequenced using Big Dye 3.1 (Applied Biosystems, Waltham, USA) chemistry and the same primers used for PCR amplification. Cycle-sequencing products were purified using Dye Ex 2.0 Spin Kit (QIAGEN, Hilden, Germany) and sequenced in an ABI3130xl Genetic analyzer (Applied Biosystems, Waltham, USA). Contig assembly of the DNA sequences was performed using the Lasergene Software package (DNASTAR, Madison, USA). The *rpoB* sequences were compared with nucleotide sequences in the GenBank database using the Basic Local Alignment Search Tool (BLAST) search algorithm. Subsequently, *rpoB* sequences were deposited on the GenBank database.

2.7. Phylogenetic Analysis

A neighbor-joining analysis was performed [43] using Molecular Evolutionary Genetics Analysis software (MEGAX) [44]. The *rpoB* gene sequences of 19 isolated strains (sequences with 100% identity with the reference sequence on BLAST were not used in the analysis), 10 *rpoB* sequences of different *Vibrio* species, and 4 genomic sequences for *Vibrio* species for which *rpoB* gene sequences were not available alone (*V. crassostreae, V. cyclitrophicus, V. celticus,* and *V. gigantis*) were considered for the phylogenetic analysis. Evolutionary distances were ascertained via the maximum composite likelihood method [45]. A bootstrap test of 1000 replicates was performed; a cut-off of 50% was set for the computation of the bootstrap condensed tree.

2.8. Statistical Analysis

Principal component analysis (PCA) was performed to reduce the dimensionality of the biochemical dataset, increasing interpretability but, at the same time, minimizing information loss. PCA was performed to illustrate the clustering of bacteria species based on biochemical features. This was accomplished considering both phenotypic (API® 20E) and molecular (PCR) species identification. Statistical analysis was performed using R software (version 1.1.463, RStudio, Inc., Boston, MA, USA).

3. Results

3.1. Necropsy, Parasitological, and Virological Examinations

A total of 20 lesser-spotted dogfish with an average weight of 2.16 ± 0.29 g (minimum weight: 1.70 g; maximum weight: 2.78 g) and an average length of 94.60 ± 4.03 mm (minimum length: 87 mm; maximum length: 103 mm) were analyzed. No external lesions were found, and the necropsy of the sharks did not show macroscopic lesions of the viscera. Parasitological (internal and external examination) and virological (Betanodavirus RT–PCR) tests yielded negative results.

3.2. Bacteriological Analysis

Out of 20 fish tested, 12 (12/20; 60%) tested positive from at least one matrix (head kidney, brain, or blood). A first visual examination of the plates was conducted, leading to the identification of 24 morphologically different isolates (Table 1). Each selected colony was subcloned to BA for subsequent tests. Following Gram staining, all isolates were identified as Gram-negative curved rods. Based on morphological and coltural characteristics, the identification by API® test envisaged the use of the API® 20E galleries. The results of the biochemical analyses carried out in a micro-method are summarized in Figure 2. On the other hand, identification by VITEK-MS did not produce any valid results according to the instrument cut-offs, so it was not possible to identify the isolates with this method.

Table 1. Characteristics of the isolates. The table shows the number of sharks, the letter linked to the different isolated colonies (A, B, C; only in specimens showing more than one colony), the matrix from which the sampling was made, and the first isolation medium. Columbia blood agar (BA); tryptic soy agar supplemented with 2% NaCl (TSA2); thiosulfate citrate bile sucrose (TCBS).

Specimen	Isolates No.	Matrix	Medium
1	A	Brain	TCBS
	B	Brain	TSA2
2		Blood	BA
3	A	Kidney	TSA2
	B	Brain	TSA2
4	A	Blood	BA
	B	Brain	TCBS
	C	Brain	TCBS
6		Kidney	TSA2
8	A	Kidney	TCBS
	B	Kidney	TCBS
9	A	Blood	TSA2
	B	Blood	TSA2
	C	Brain	TCBS
11		Brain	TCBS
13	A	Kidney	TSA2
	B	Blood	TSA2
	C	Brain	TSA2
14	A	Blood	TSA2
	B	Blood	TSA2
19	A	Kidney	TSA2
	B	Brain	BA
20	A	Kidney	TSA2
	B	Blood	TSA2

Isolate	ONPG	ADH	LDC	ODC	CIT	H₂S	URE	TDA	IND	VP	GEL	GLU	MAN	INO	SOR	RHA	SAC	MEL	AMY	ARA	OX	ID Api	ID gen
1A	-	+	-	-	-	-	-	-	+	+	+	+	+	-	-	-	-	-	+	-	+	*Vibrio celticus*	*V. crossostreae*
1B	-	-	-	-	-	-	-	-	+	-	+	+	+	-	-	-	-	-	+	-	+	*V. gallicus*	*V. crossostreae*
2	-	+	-	-	-	-	-	-	+	+	+	+	+	-	-	-	-	-	+	-	+	*V. celticus*	*V. crossostreae*
3A	+	+	-	-	-	-	-	-	+	+	+	+	+	-	-	-	-	+	+	-	+	*Aeromonas hydrophila*	*V. cyclitrophicus*
3B	+	+	-	-	-	-	-	-	+	+	+	+	+	-	-	-	-	+	+	-	+	*A. hydrophila*	*V. cyclitrophicus*
4A	-	+	-	-	-	-	-	-	+	-	+	+	+	-	-	-	-	+	+	-	+	*V. gigantis*	*V. cyclitrophicus*
4B	+	+	-	-	-	-	-	-	+	-	+	+	+	-	-	-	-	+	+	-	+	*V. splendidus*	*V. crossostreae*
4C	-	-	-	-	-	-	-	-	+	-	+	+	+	-	-	-	-	+	+	-	+	*V. gallicus*	*V. cyclitrophicus*
6	+	+	-	-	-	-	-	-	+	+	+	+	+	-	-	+	-	+	+	-	+	*A. hydrophila*	*V. crossostreae*
8A	+	+	-	-	-	-	-	-	+	-	+	+	+	-	-	-	-	+	+	-	+	*V. splendidus*	*V. crossostreae*
8B	+	+	-	-	-	-	-	-	+	-	+	+	+	-	-	-	-	+	+	-	+	*V. splendidus*	*V. crossostreae*
9A	-	+	-	-	-	-	-	-	+	+	+	+	+	-	-	-	-	-	+	-	+	*V. celticus*	*V. crossostreae*
9B	-	-	-	-	-	-	-	-	+	-	+	+	+	-	-	-	-	-	+	-	+	*V. gallicus*	*V. cyclitrophicus*
9C	-	+	-	-	-	-	-	-	+	+	+	+	+	-	-	-	-	-	+	-	+	*V. celticus*	*V. crossostreae*
11	-	+	-	-	-	-	-	-	+	+	+	+	+	-	-	-	-	-	+	-	+	*V. celticus*	*V. crossostreae*
13A	-	+	-	-	-	-	-	-	+	-	+	+	+	-	-	-	-	+	+	-	+	*V. gigantis*	*V. cyclitrophicus*
13B	+	+	-	-	-	-	-	+	+	-	+	+	+	-	-	-	+	-	+	-	+	*V. fluvialis*	*V. cyclitrophicus*
13C	+	+	-	-	-	-	-	+	+	-	+	+	+	-	-	-	+	-	+	-	+	*V. fluvialis*	*V. cyclitrophicus*
14A	+	+	-	-	+	-	-	-	+	+	+	+	+	-	-	-	-	+	+	-	+	*A. hydrophila*	*V. crossostreae*
14B	+	+	-	-	+	-	-	-	+	+	+	+	+	-	-	-	-	+	+	-	+	*A. hydrophila*	*V. crossostreae*
19A	+	+	-	-	-	-	-	-	+	-	+	+	+	-	-	-	-	+	+	-	+	*A. hydrophila*	*V. cyclitrophicus*
19B	+	+	-	-	-	-	-	-	+	-	+	-	-	-	-	-	-	-	+	-	+	*V. fluvialis*	*V. crossostreae*
20A	-	+	-	-	-	-	-	-	+	+	+	+	+	-	-	-	-	+	+	-	+	*V. gigantis*	*V. cyclitrophicus*
20B	-	+	-	-	-	-	-	-	+	-	+	+	+	-	-	-	-	+	+	-	+	*V. gigantis*	*V. cyclitrophicus*

Figure 2. Biochemical features of isolates (API® 20E). ONPG: test for β-galactosidase enzyme by hydrolysis of the substrate o-nitrophenyl-b-D-galactopyranoside; ADH, decarboxylation of the amino acid arginine by arginine dihydrolase; LDC, decarboxylation of the amino acid lysine by lysine decarboxylase; ODC, decarboxylation of the amino acid ornithine by ornithine decarboxylase; CIT, utilization of citrate as only carbon source; H₂S, production of hydrogen sulfide; URE, test for the enzyme urease; TDA, tryptophan deaminase; detection of the enzyme tryptophan deaminase: reagent, ferric chloride; IND: indole test production of indole from tryptophan by the enzyme tryptophanase. Reagent–indole is detected by addition of Kovac's reagent; VP, the Voges–Proskauer test for the detection of acetoin (acetyl methylcarbinol) produced by fermentation of glucose by bacteria utilizing the butylene glycol pathway; GEL, test for the production of the enzyme gelatinase, which liquefies gelatin; GLU, fermentation of glucose (hexose sugar); MAN, fermentation of mannose (hexose sugar); INO, fermentation of inositol (cyclic polyalcohol); SOR, fermentation of sorbitol (alcohol sugar); RHA, fermentation of rhamnose (methyl pentose sugar); SAC, fermentation of sucrose (disaccharide); MEL, fermentation of melibiose (disaccharide); AMY, fermentation of amygdalin (glycoside); ARA, fermentation of arabinose (pentose sugar). In the last two columns, there is the API identifier and the real identifier provided by the molecular analysis.

3.3. Biomolecular Analysis for Bacterial Identification

All the 24 strains isolated showed a fragment of 730 bp from the *rpoB* gene amplification. Of those, 11 isolates (11/24; 45.8%) were identified at species level as *Vibrio cyclitrophicus* and 13 (13/24; 54.2%) as *V. crossostreae* (BLASTn nucleotide sequence identity value ranging from 98 to 100%). Results are reported in Table 2.

All obtained sequences, except for sequences with 100% identity value with the reference sequences in BLAST, were deposited on GenBank, a freely available online database (Accession Numbers: OM158211-OM158229).

Table 2. Isolates identifier, identification (ID), percentage (%) of identity compared to reference sequences, and accession numbers of sequences deposited on GenBank. *: the sequences with an identity percentage equal to 100% to reference sequences already deposited on GenBank were not submitted.

Isolates No.	Genetic ID	ID (%)	Reference for Identity	GenBank Accession
1A	*V. crossostreae*	98.71	CP016228	OM158222
1B	*V. crossostreae*	98.85	CP016228	OM158223
2	*V. crossostreae*	98.69	CP016228	OM158216
3A	*V. cyclitrophicus*	99.86	CP039700	OM158211
3B	*V. cyclitrophicus*	100	CP039700	not submitted *
4A	*V. cyclitrophicus*	99.85	CP039700	OM158217
4B	*V. crossostreae*	98.66	CP016228	OM158224

Table 2. *Cont.*

Isolates No.	Genetic ID	ID (%)	Reference for Identity	GenBank Accession
4C	*V. cyclitrophicus*	99.86	CP039700	OM158225
6	*V. crassostreae*	98.8	CP016228	OM158212
8A	*V. crassostreae*	98.81	CP016228	OM158213
8B	*V. crassostreae*	98.82	CP016228	OM158214
9A	*V. crassostreae*	98.71	CP016228	OM158218
9B	*V. cyclitrophicus*	100	CP039700	not submitted *
9C	*V. crassostreae*	98.71	CP016228	OM158226
11	*V. crassostreae*	98.71	CP016228	OM158227
13A	*V. cyclitrophicus*	100	CP039700	not submitted *
13B	*V. cyclitrophicus*	100	CP039700	not submitted *
13C	*V. cyclitrophicus*	99.86	CP039700	OM158228
14A	*V. crassostreae*	99.2	CP016228	OM158219
14B	*V. crassostreae*	98.72	CP016228	OM158220
19A	*V. cyclitrophicus*	100	CP039700	not submitted *
19B	*V. crassostreae*	98.72	CP016228	OM158229
20A	*V. cyclitrophicus*	99.86	CP039700	OM158215
20B	*V. cyclitrophicus*	99.86	CP039700	OM158221

3.4. Phylogenetic Analysis

The bootstrap condensed tree is shown in Figure 3. Two main clusters are present: one containing the 10 *rpoB* sequences of different *Vibrio* species subdivided into three different subclusters, the second containing *V. cyclitrophicus* sequences with a high bootstrap value (97%). *V. crassostreae* isolates are not shown to be clustering with other *Vibrio* species, considering a bootstrap value cut-off of 50%.

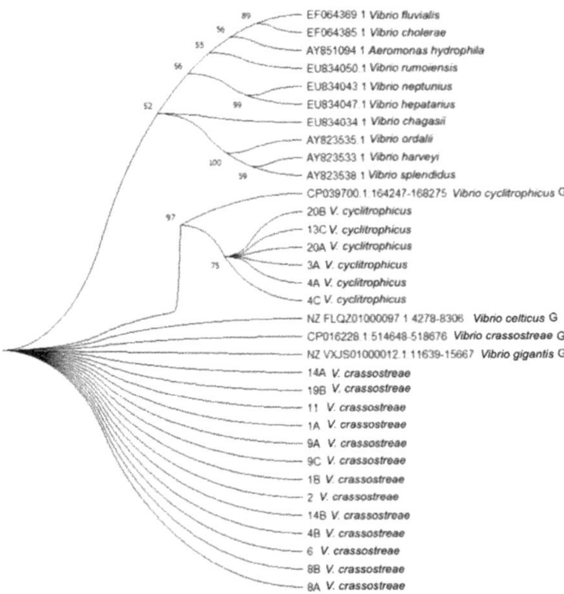

Figure 3. Phylogenetic relationship among *Vibrio* species isolated from lesser-spotted dogfish and different *Vibrio* species. Phylogenetic tree was constructed using MEGAX and neighbor-joining method. A bootstrap test of 1000 replicates was performed; the bootstrap condensed tree, using a cut-off value of 50% is shown. The sequences with "G" displayed after the scientific name derived from a deposited genome.

3.5. Biochemical Features of Isolates

Morphological and biochemical analysis showed that all isolates were Gram-negative bacteria, positive for cytochrome oxidase, produced indole from tryptophan, and liquefied gelatin from enzyme gelatinase. None of the isolated strains produced lysine decarboxylase, ornithine decarboxylase, hydrogen sulfide, urease, nor were they able to utilize citrate. The enzyme tryptophan deaminase was negative in almost all strains. Generally, the isolates produced acid from glucose, arabinose, amygdalin, and mannitol. No acid was produced from amygdalin and arabinose. It is important to highlight a certain variability in biochemical features for the following test: decarboxylation of the amino acid arginine by arginine dihydrolase, detection of acetoin (acetyl methylcarbinol) produced by fermentation of glucose by bacteria utilizing the butylene glycol pathway (Voges–Proskauer test), and fermentation of sucrose and melibiose (Figure 2). The first PCA (Figure 4) based on phenotypic identification (API® 20E) showed that the first (PC1) and second (PC2) components accounted for meaningful amounts of the total variance (87.96%). PC1 explained 78.1% of the total variance, and PC2, 9.86%. PCA of bacteria strains yielded three main clusters. The blue cluster contains most of the bacteria species (*Vibrio celticus*, *V. gigantis*, and *V. gallium*) that were similar in biochemical features. The red cluster contains only *V. fluvialis*. Finally, the green cluster contains only *Aeromonas hydrophila* and *V. splendidus*. The second PCA cluster analysis (Figure 5) based on molecular species identification showed that the first (PC1) and second (PC2) components accounted for meaningful amounts of the total variance (58.93%). PC1 explained 30.86% of the total variance, and PC2, 28.07%. PCA of bacteria strains yielded two main clusters. The red cluster contains most of the strains that belonged to *V. crassostreae*, whereas the blue cluster contains *V. cyclitrophicus*. It is possible to observe a partial overlap of the two bacteria clusters, confirmed by the overlap of the confidence ellipses (95%) of each one, indicating similar biochemical features for certain tests.

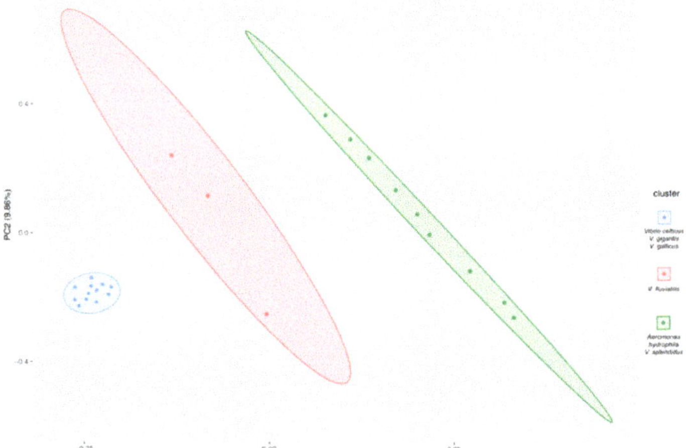

Figure 4. Principal component analysis of biochemical features of *Vibrio* spp. strains based on phenotypic species identification (API® 20E test). The blue cluster contains most of the strains that were similar in biochemical features (*Vibrio celticus*, *V. gigantis*, and *V. gallicus*). The red cluster contains *V. fluvialis*, whereas the green cluster contains *Aeromonas hydrophila* and *V. splendidus*. Confidence ellipses (95%) plot convex hull values of each cluster.

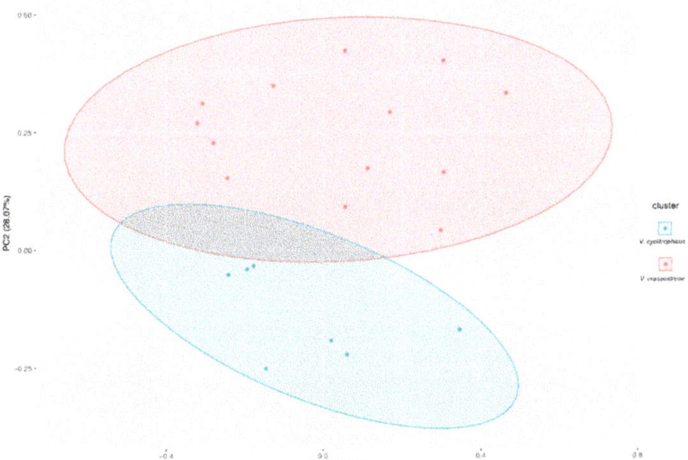

Figure 5. Principal component analysis of biochemical features of *Vibrio* spp. strains based on molecular species identification. The red cluster contains most of the strains that belonged to *V. crassostreae*, whereas the blue cluster contains *V. cyclitrophicus*. Confidence ellipses (95%) plot convex hull values of each cluster.

4. Discussion

This study reports the finding of *Vibrio crassostreae* and *V. cyclotrophicus* in *Scyliorhinus canicula* isolated from different matrices (brain, blood, and kidney) and provides a description of the isolated strains. Notably, there are only a few reports in the literature on the isolation of pathogens from sharks. Therefore, our findings could clarify the pathogenic role of these bacteria despite the absence of gross pathology lesions.

The isolation process of *Vibrio* spp. from sharks started following the observation of external clinical signs common to many infectious diseases. For this reason, the presence of parasites, bacteria, and viruses was investigated. Parasitological and virological analyses tested negative, although most cases of diseases in sharks reported in the literature were of parasitic etiology [37]. In relation to the symptoms highlighted in the tank, a possible viral cause different from Betanodavirus cannot be excluded. However, it was not possible to carry out in-depth analyses due to the low number of samples, the limited amount of biological material available (in relation to shark size), and the scarce presence of information regarding elasmobranch viruses.

On the contrary, colonial growth was observed on agar media. Once pure cultures of the morphologically different colonies were obtained, the diagnostic techniques available in the laboratory clearly identified the isolates. Therefore, the isolates were tested by a biochemical test (API® 20E), MALDI-TOF MS, and molecular biology techniques. Upon Gram staining, the isolates were all found to be Gram-negative rods. Since the colonies grew on TCBS and had the typical morphology of *Vibrio* spp., it was decided to use the API® 20E galleries for species identification and characterization based on a biochemical test.

Biochemical tests led to the identification of bacteria of the genus *Vibrio*, as presumed, but also of *Aeromonas hydrophila* isolates. Therefore, the identification process continued by VITEK-MS, which did not lead to the identification of any of the isolates. The possible explanation of this was probably due to the absence in the database used by the instrument of reference strains comparable to those of the study. The implementation of these databases with *Vibrio* spp. strains not belonging from food or clinical isolates could make MALDI-TOF MS a rapid and efficient method for diagnosing these bacteria. Moreover, the identification of strains identified as *A. hydrophila* by biochemical methods was not confirmed, although it is reported that VITEK-MS is able to distinguish *Aeromonas* spp. from some species of

Vibrio (*V. cholerae*) [46]. For this reason, it was decided to use molecular biology techniques to obtain a more accurate identification.

For genetic identification, the 16S subunit of ribosomal RNA (16S rRNA) gene is commonly used for the study of phylogenetic relationship among bacterial taxa and identification at the species level; however, the 16S rRNA genes of vibrios are not sufficiently polymorphic to ensure a reliable identification [47–49]. For this reason, species-specific PCRs have been developed, and different genetic markers have been considered. Some examples are the gene encoding the α-subunit of bacterial ATP synthase (*atpA*) [50] and *rpoB* [42]; considering that the *rpoB* gene led to more suitable results than the 16S rRNA gene for *Vibrio* species identification; thus, this tool was chosen in this study for the molecular identification. The mentioned analysis allowed the identification of the isolates, respectively, as *V. crassostreae* (13/24; 54.2%) and *V. cyclitrophicus* (11/24; 45.8%). The first species was reported as a pathogen of mollusks and fish [51], while *V. cyclitrophicus* was reported in mollusks and in the microbiome of marine copepods [52,53]. There are no reports of these species in sharks, although bacteria of the genus *Vibrio* have already been reported in elasmobranchs but not specifically for *S. canicula*.

As further support of these results, a statistical analysis was carried out on the results of biochemical and molecular tests. The biochemical test (API® 20E) allowed the identification of three main clusters of *Vibrio* species, based on biochemical features. In particular, the first PCA revealed that strains belonging to different *Vibrio* species (*V. celticus, V. gigantis,* and *V. gallicus*) had the same biochemical features. Considering this, it is clear how such identification is not a good diagnostic tool in fish pathology. On the contrary, the second PCA performed on biochemical features of *Vibrio* species identified throughout molecular tools allowed the discrimination of the biochemical features of *V. crassostreae* and *V. cyclotrophicus*, even though a partial overlap of the two main clusters was observed, indicating similar features for a certain aspect of the biochemical test (e.g., negative for decarboxylations of the amino acid ornithine, decarboxylation of the amino acid lysine urease, H_2S production). These results support what has already been indicated in the description of the two *Vibrio* species isolated in this survey [54,55].

The phylogenetic analysis showed a clear separation between *V. cyclitrophicus*, characterized by a cluster with a high bootstrap value (97%), and *V. crassostreae*, even if a partial overlap was detected in the PCA analysis.

Similarly, *Aeromonas hydrophila, V. splendidus,* and *V. fluvialis* in the phylogenetic tree were clearly separated from *V. crassostreae* and *V. cyclitrophicus* even if sometimes identified as them by the API system.

Vibrio species identified in the study are generally linked to mortality in bivalve mollusks or found at the environmental level. A study by Petton et al. [56] shows how *Vibrio* infection is related to water temperature: In this article, it is shown that temperatures higher or lower than 16 °C affect oyster mortality due to the growth of *Vibrio* spp. As indicated by Sims [57], *S. canicula* fits perfectly in a thermal range between 14.9 °C and 17.7 °C. Thus, the tank's water temperature is ideal both for the growth of isolated *Vibrio* species and for *S. canicula*. Nevertheless, it has been found that in natural environments, the lesser-spotted dogfish prefers temperatures lower than 17 °C, for optimization of its metabolism [57].

Currently, few studies are available in the literature on elasmobranchs infectious diseases. Probably, these shortcomings are due to the lack of general elasmobranch studies, as supported by Abdul Malak et al. [32]. Among bacterial diseases, *Vibrio* spp. infections are reported, although in a few reports and with several shortcomings in the identification process. The present study, therefore, has the function of a report of the presence of vibrios, similar to previous research; in-depth studies will be required for details on the pathogenic mechanism of these bacteria in sharks. Nevertheless, our study made it possible to identify specifically *V. crassostreae* and *V. cyclitrophicus*, underlining the criticalities in the identification process in relation to the techniques adopted (API® 20E, MALDI-TOF MS, molecular biology) and the need for developing specific guidelines for the identification of nonmajor pathogenic *Vibrio* species.

Author Contributions: Conceptualization: M.T., G.E., D.M., P.P., M.P. and S.C.; methodology: M.T., G.E., D.M., P.P., M.P. and S.C.; investigation: M.T., G.E., N.S., V.M., F.G., I.C., A.P., M.P. and S.C.; resources: P.L.A., A.D. and M.P.; data curation: M.T., G.E., P.P., M.P. and S.C.; writing—original draft preparation: M.T. and G.E.; writing—review and editing: D.M., P.P., M.P. and S.C.; visualization: M.T., G.E., D.M., P.P., M.P. and S.C.; supervision: P.L.A., A.D. and M.P. All authors have read and agreed to the published version of the manuscript.

Funding: This research received no external funding.

Institutional Review Board Statement: The Ethics Committee was not included, as all samples were derived from a diagnostic service that the laboratory in which the authors work offers to users (http://www.izsplv.it/) Therefore, the study was derived from a routine activity for which it was not necessary to establish an Ethics Committee.

Informed Consent Statement: Not applicable.

Data Availability Statement: Not applicable.

Conflicts of Interest: The authors declare no conflict of interest.

References

1. Reen, F.J.; Almagro-Moreno, S.; Ussery, D.; Boyd, E.F. The genomic code inferring *Vibrionaceae* niche specialization. *Nat. Rev. Microbiol.* **2006**, *9*, 697–704. [CrossRef] [PubMed]
2. Johnson, C.N. Fitness factors in vibrios: A mini-review. *Microb. Ecol.* **2013**, *65*, 826–851. [CrossRef] [PubMed]
3. Parte, A.C.; Sardà Carbasse, J.; Meier-Kolthoff, J.P.; Reimer, L.C.; Göker, M. List of Prokaryotic names with Standing in Nomenclature (LPSN) moves to the DSMZ. *Int. J. Syst. Evol.* **2020**, *70*, 5607–5612. [CrossRef] [PubMed]
4. Colwell, R.R.; Soira, W.M. The ecology of *Vibrio Cholera*. In *Cholera*; Barua, D., Greenough, W.B., Eds.; Springer: Boston, MA, USA, 1992; pp. 1–36, ISBN 978-1-4757-9690-2.
5. Sorum, H.; Myhr, E.; Zwicker, B.M.; Lillehaug, A. Comparison by plasmid profiling of *Vibrio salmonicida* strains isolated from diseased fish from different north European and Canadian coastal areas of the Atlantic Ocean. *Can. J. Fish Aquat. Sci.* **1993**, *50*, 247–250. [CrossRef]
6. Diggles, B.K.; Carson, J.; Hine, P.M.; Hickman, R.W.; Tait, M.J. *Vibrio* species associated with mortalities in hatchery-reared turbot (*Colistium nudipinnis*) and brill (*C. guntheri*) in New Zealand. *Aquaculture* **2000**, *183*, 1–12. [CrossRef]
7. Goarant, C.; Herlin, J.; Brizard, R.; Marteau, A.L.; Martin, C.; Martin, B. Toxic factors of *Vibrio* strains pathogenic to shrimp. *Dis. Aquat. Org.* **2000**, *40*, 101–107. [CrossRef] [PubMed]
8. Takahashi, K.G.; Nakamura, A.; Mori, K. Inhibitory effects of ovoglobulins on bacillary necrosis in larvae of the Pacific oyster, *Crassostrea gigas*. *J. Invertebr. Pathol.* **2000**, *75*, 212–217. [CrossRef] [PubMed]
9. Paillard, C.; Le Roux, F.; Borrego, J.J. Bacterial disease in marine bivalves, a review of recent studies: Trends and evolution. *Aquat. Living Resour.* **2004**, *17*, 477–498. [CrossRef]
10. Liu, P.C.; Lin, J.Y.; Hsiao, P.T.; Lee, K.K. Isolation and characterization of pathogenic *Vibrio alginolyticus* from diseased cobia *Rachycentron canadum*. *J. Basic Microbiol.* **2004**, *44*, 23–28. [CrossRef] [PubMed]
11. Zhao, D.H.; Sun, J.J.; Liu, L.; Zhao, H.H.; Wang, H.F.; Liang, L.Q.; Liu, L.B.; Li, G.F. Characterization of two phenotypes of *Photobacterium damselae* subsp. *damselae* isolated from diseased juvenile *Trachinotus ovatus* reared in cage mariculture. *J. World Aquacult. Soc.* **2009**, *40*, 281–289. [CrossRef]
12. Zhang, X.; Li, Y.W.; Mo, Z.Q.; Luo, X.C.; Sun, H.Y.; Liu, P.; Li, A.X.; Zhou, S.M.; Dan, X.M. Outbreak of a novel disease associated with *Vibrio mimicus* infection in fresh water cultured yellow catfish, *Pelteobagrus fulvidraco*. *Aquaculture* **2014**, *432*, 119–124. [CrossRef]
13. Rajan, P.R.; Lopez, C.; Lin, J.H.; Yang, H. *Vibrio alginolyticus* infection in cobia (*Rachycentron canadum*) cultured in Taiwan. *Bull. Eur. Ass. Fish Pathol.* **2001**, *21*, 228–234.
14. Ransangan, J.; Mustafa, S. Identification of *Vibrio harveyi* isolated from diseased Asian seabass (*Lates calcarifer*) by use of 16S ribosomal DNA sequencing. *J. Aquat. Anim. Health* **2009**, *21*, 150–155. [CrossRef] [PubMed]
15. Austin, B.; Austin, D. *Bacterial Fish Pathogens: Diseases of Farmed and Wild Fish*, 6th ed.; Springer International Publishing: Dordrecht, The Netherlands, 2016.
16. Akayli, T.; Timur, G.; Albayrak, G.; Aydemir, B. Identification and genotyping of *Vibrio ordalii*: A comparison of different methods. *Isr. J. Aquacult-Bamid* **2010**, *62*, 9–18. [CrossRef]
17. Dong, H.T.; Taengphu, S.; Sangsuriya, P.; Charoensapsri, W.; Phiwsaiya, K.; Sornwatana, T.; Khunrae, P.; Rattanarojpong, T.; Senapin, S. Recovery of *Vibrio harveyi* from scale drop and muscle necrosis disease in farmed barramundi, *Lates calcarifer* in Vietnam. *Aquaculture* **2017**, *473*, 89–96. [CrossRef]

18. Labella, A.; Vida, M.; Alonso, M.C.; Infante, C.; Cardenas, S.; Lopez–Romalde, S.; Manchado, M.; Borrego, J.J. First isolation of *Photobacterium damselae* ssp. *damselae* from cultured redbanded seabream, *Pagrus auriga* Valenciennes, in Spain. *J. Fish Dis.* **2006**, *2*, 175–179. [CrossRef]

19. Nurliyana, M.; Amal, M.N.A.; Zamri-Saad, M.; Ina-Salwany, M.Y. Possible transmission routes of *Vibrio* spp. in tropical cage-cultured marine fishes. *Lett. Appl. Microbiol.* **2019**, *68*, 485–496. [CrossRef]

20. Kim, M.N.; Bang, H.J. Detection of marine pathogenic bacterial *Vibrio* species by multiplex polymerase chain reaction (PCR). *J. Environ. Biol.* **2008**, *29*, 543–546.

21. Sandlund, N.; Rødseth, O.M.; Knappskog, D.H.; Fiksdal, I.U.; Bergh, Ø. Comparative susceptibility of turbot, halibut, and cod yolk-sac larvae to challenge with *Vibrio* spp. *Dis. Aquat. Org.* **2010**, *89*, 29–37. [CrossRef]

22. Haenen, O.L.M.; Van Zanten, E.; Jansen, R.; Roozenburg, I.; Engelsma, M.Y.; Dijkstra, A.; Möller, A.V.M. *Vibrio vulnificus* outbreaks in Dutch eel farms since 1996: Strain diversity and impact. *Dis. Aquat. Org.* **2014**, *108*, 201–209. [CrossRef] [PubMed]

23. Bellos, G.; Angelidis, P.; Miliou, H. Effect of temperature and seasonality principal epizootiological risk factor on vibriosis and photobacteriosis outbreaks for european sea bass in greece (1998–2013). *J. Aquac. Res. Dev.* **2015**, *6*, 10–4172. [CrossRef]

24. Liu, C.H.; Wu, K.; Chu, T.W.; Wu, T.M. Dietary supplementation of probiotic, *Bacillus subtilis* E20, enhances the growth performance and disease resistance against *Vibrio alginolyticus* in parrot fish (*Oplegnathus fasciatus*). *Aquac. Int.* **2018**, *26*, 63–74. [CrossRef]

25. Grimes, D.J.; Colwell, R.R.; Stemmler, J.; Hada, H.; Maneval, D.; Hetrick, F.M.; Stoskopf, M. *Vibrio* species as agents of elasmobranch disease. *Helgol. Meeresunters.* **1984**, *37*, 309–315. [CrossRef]

26. Heng, S.P.; Letchumanan, V.; Deng, C.Y.; Ab Mutalib, N.S.; Khan, T.M.; Chuah, L.H.; Chan, K.G.; Goh, B.H.; Pusparajah, P.; Lee, L.H. *Vibrio vulnificus*: An environmental and clinical burden. *Front. Microbiol.* **2017**, *8*, 997. [CrossRef]

27. Libralato, S.; Christensen, V.; Pauly, D. A method for identifying keystone species in food web models. *Ecol. Modell.* **2006**, *195*, 153–171. [CrossRef]

28. Baum, J.K.; Worm, B. Cascading top-down effects of changing oceanic predator abundances. *J. Anim. Ecol.* **2009**, *78*, 699–714. [CrossRef]

29. Ferretti, F.; Myers, R.A.; Serena, F.; Lotze, H.K. Loss of large predatory sharks from the Mediterranean Sea. *Biol. Conserv.* **2008**, *22*, 952–964. [CrossRef] [PubMed]

30. Coll, M.; Piroddi, C.; Steenbeek, J.; Kaschner, K.; Ben Rais Lasram, F.; Aguzzi, J.; Voultsiadou, E. The biodiversity of the Mediterranean Sea: Estimates, patterns, and threats. *PLoS ONE* **2010**, *5*, e11842. [CrossRef] [PubMed]

31. Coll, M.; Navarro, J.; Palomera, I. Ecological role, fishing impact, and management options for the recovery of a Mediterranean endemic skate by means of food web models. *Biol. Conserv.* **2013**, *157*, 108–120. [CrossRef]

32. Malak, D.A. *Overview of the Conservation Status of the Marine Fishes of the Mediterranean Sea*; IUCN: Gland, Switzerland, 2011.

33. Lyle, J.M. Food and feeding habits of the lesser spotted dogfish, *Scyliorhinus canicula* (L.) in Isle of Man waters. *J. Fish. Biol.* **1983**, *23*, 139–148. [CrossRef]

34. Olaso, I.; Velasco, F.; Pérez, N. Importance of discarded blue whiting (*Micromessistius poutassou*) in the diet of lesser spotted dogfish (*Scyliorhinus canicula*) in the Cantabrian Sea. *ICES J. Mar. Sci.* **1998**, *55*, 331–341. [CrossRef]

35. Lee, K.A.; Huveneers, C.; Peddemors, V.; Boomer, A.; Harcourt, R.G. Born to be free? Assessing the viability of releasing captive-bred wobbegongs to restock depleted populations. *Front. Mar. Sci.* **2015**, *2*, 18. [CrossRef]

36. Terrell, S.P. An Introduction to Viral, Bacterial and Fungal Diseases of Elasmobranchs. In *Elasmobranch Husbandry Manual: Captive Care of Sharks, Rays and Their Relatives*; Smith, M., Warmolts, D., Thoney, D., Hueter, R., Eds.; Ohio Biological Survey Inc.: Columbus, OH, USA, 2004; p. 589.

37. Bakopoulos, V.; Tsepa, E.; Diakou, A.; Kokkoris, G.; Kolygas, M.; Athanassopoulou, F. Parasites of *Scyliorhinus canicula* (Linnaeus, 1758) in the north-eastern Aegean Sea. *J. Mar. Biolog. Assoc.* **2018**, *98*, 2133–2143. [CrossRef]

38. Gratzek, J.B.; Gilbert, J.P.; Lohr, A.L.; Shotts, E.B., Jr.; Brown, J. Ultraviolet light control of *Ichthyophthirius multifiliis* Fouquet in a closed fish culture recirculation system. *J. Fish Dis.* **1983**, *6*, 145–153. [CrossRef]

39. Pratesi, R. L'interazione radiazione ottica-biomateria: Meccanismi d'azione, danni biologici e cautele d'uso. In *Quaderni di Tecniche di Protezione Ambientale a Cura di Adriano Zavatti*; Pitagora Editrice: Bologna, Italy, 1993.

40. Panzarin, V.; Patarnello, P.; Mori, A.; Rampazzo, E.; Cappellozza, E.; Bovo, G.; Cattoli, G. Development and validation of a real-time TaqMan PCR assay for the detection of betanodavirus in clinical specimens. *Arch. Virol.* **2010**, *155*, 1193–1203. [CrossRef]

41. Pastorino, P.; Colussi, S.; Pizzul, E.; Varello, K.; Menconi, V.; Mugetti, D.; Tomasoni, M.; Esposito, G.; Bertoli, M.; Bozzetta, E.; et al. The unusual isolation of carnobacteria in eyes of healthy salmonids in high-mountain lakes. *Sci. Rep.* **2021**, *11*, 2314. [CrossRef] [PubMed]

42. Ki, J.S.; Zhang, R.; Zhang, W.; Huang, Y.L.; Qian, P.Y. Analysis of RNA polymerase beta subunit (*rpoB*) gene sequences for the discriminative power of marine *Vibrio* species. *Microb. Ecol.* **2009**, *58*, 679–691. [CrossRef]

43. Saitou, N.; Nei, M. The neighbor-joining method: A new method for reconstructing phylogenetic trees. *Mol. Biol. Evol.* **1987**, *4*, 406–425. [CrossRef] [PubMed]

44. Kumar, S.; Stecher, G.; Li, M.; Knyaz, C.; Tamura, K. MEGA X: Molecular Evolutionary Genetics Analysis across computing platforms. *Mol. Biol. Evol.* **2018**, *35*, 1547–1549. [CrossRef] [PubMed]

45. Tamura, K.; Nei, M.; Kumar, S. Prospects for Inferring Very Large Phylogenies by Using the Neighbor-Joining Method. *Proc. Natl. Acad. Sci. USA* **2004**, *101*, 11030–11035. [CrossRef]

46. Rychert, J.; Creely, D.; Mayo-Smith, L.M.; Calderwood, S.B.; Ivers, L.C.; Ryan, E.T.; Boncy, J.; Qadri, F.; Ahmed, D.; Ferraro, M.J.; et al. Evaluation of matrix-assisted laser desorption ionization-time of flight mass spectrometry for identification of *Vibrio cholerae*. *J. Clin. Microbiol.* **2015**, *53*, 329–331. [CrossRef]

47. Tarr, C.L.; Patel, J.S.; Puhr, N.D.; Sowers, E.G.; Bopp, C.A.; Strockbine, N.A. Identification of *Vibrio* isolates by a multiplex PCR assay and *rpoB* sequence determination. *J. Clin. Microbiol.* **2007**, *45*, 134–140. [CrossRef] [PubMed]

48. Chun, J.; Huq, A.; Colwell, R.R. Analysis of 16S–23S rRNA intergenic spacer regions of *Vibrio cholerae* and *Vibrio mimicus*. *Appl. Environ. Microbiol.* **1999**, *65*, 2202–2208. [CrossRef] [PubMed]

49. Thompson, F.L.; Gevers, D.; Thompson, C.C.; Dawyndt, P.; Naser, S.; Hoste, B.; Munn, C.B.; Swings, J. Phylogeny and molecular identification of vibrios on the basis of multilocus sequence analysis. *Appl. Environ. Microbiol.* **2005**, *71*, 5107–5115. [CrossRef]

50. Thompson, C.C.; Thompson, F.L.; Vicente, A.C.P.; Swings, J. Phylogenetic analysis of vibrios and related species by means of *atpA* gene sequences. *Int. J. Syst. Evol. Microbiol.* **2007**, *57*, 2480–2484. [CrossRef] [PubMed]

51. Sohn, H.; Kim, J.; Jin, C.; Lee, J. Identification of *Vibrio* species isolated from cultured olive flounder (*Paralichthys olivaceus*) in Jeju Island, South Korea. *Fish. Aquat. Sci.* **2019**, *22*, 14. [CrossRef]

52. Nuttall, R.; Sharma, G.; Moisander, P.H. Draft Genome Sequence of *Vibrio cyclitrophicus* NCT10V, Cultivated from the Microbiome of a Marine Copepod. *Microbiol. Resour. Announc.* **2019**, *8*, e01208-19. [CrossRef]

53. Li, Y.F.; Chen, Y.W.; Xu, J.K.; Ding, W.Y.; Shao, A.Q.; Zhu, Y.T.; Yang, J.L. Temperature elevation and *Vibrio cyclitrophicus* infection reduce the diversity of haemolymph microbiome of the mussel *Mytilus coruscus*. *Sci. Rep.* **2019**, *9*, 1–10. [CrossRef]

54. Hedlund, B.P.; Staley, J.T. *Vibrio cyclotrophicus* sp. nov., a polycyclic aromatic hydrocarbon (PAH)-degrading marine bacterium. *Int. J. Syst. Evol. Microbiol.* **2001**, *51*, 61–66. [CrossRef]

55. Faury, N.; Saulnier, D.; Thompson, F.L.; Gay, M.; Swings, J.; Roux, F.L. *Vibrio crassostreae* sp. nov., isolated from the haemolymph of oysters (*Crassostrea gigas*). *Int. J. Syst. Evol. Microbiol.* **2004**, *54*, 2137–2140. [CrossRef]

56. Petton, B.; Boudry, P.; Alunno-Bruscia, M.; Pernet, F. Factors influencing disease-induced mortality of Pacific oysters *Crassostrea gigas*. *Aquac. Environ. Interact.* **2015**, *6*, 205–222. [CrossRef]

57. Sims, D.W. Tractable models for testing theories about natural strategies: Foraging behaviour and habitat selection of free-ranging sharks. *J. Fish Biol.* **2003**, *63*, 53–73. [CrossRef]

Journal of
Marine Science and Engineering

MDPI

Article

Skin Culturable Microbiota in Farmed European Seabass (*Dicentrarchus labrax*) in Two Aquacultures with and without Antibiotic Use

Ana Ramljak [1], Irena Vardić Smrzlić [1], Damir Kapetanović [1,*], Fran Barac [1], Anamarija Kolda [1], Lorena Perić [1], Ivana Balenović [2], Tin Klanjšček [1] and Ana Gavrilović [3]

[1]   Ruđer Bošković Institute, 10000 Zagreb, Croatia; ana.ramljak@irb.hr (A.R.); ivardic@irb.hr (I.V.S.); fran.barac.3@gmail.com (F.B.); anamarija.kolda@irb.hr (A.K.); lorena.peric@irb.hr (L.P.); tin.klanjscek@irb.hr (T.K.)
[2]   Orada Adriatic d.o.o., 51557 Cres, Croatia; ivana.balenovic@orada-adriatic.hr
[3]   Faculty of Agriculture, University of Zagreb, 10000 Zagreb, Croatia; agavrilovic@agr.hr
*   Correspondence: kada@irb.hr

Abstract: This study examined culturable skin microbiota that was associated with farmed European seabass (*Dicentrarchus labrax*). Healthy European seabass were sampled during summer commercial harvest from one conventional fish farm where antibiotics are used, and from another practicing a certified antibiotic-free fish aquaculture. Physicochemical and microbiological analysis of seawater and sediment were performed, as well as determination of culturable bacteria, including *Vibrio*, from skin swabs of European seabass and seawater and sediment at both farms. Samples were processed for isolation of bacteria and their characterization by molecular and antibiotic susceptibility tests. In both fish farms, most of the bacteria that were identified in the skin belonged to the genera *Pseudomonas* and *Vibrio*. Some of the microbiota that were identified are known to be pathogenic to fish: *V. alginolyticus*, *V. anguillarum*, and *V. harveyi*. *Vibrio* strains showed higher resistance to certain antibiotics compared to previous studies. This study provides, for the first time, information on the culturable skin bacteria that is associated with healthy European seabass under culture conditions with and without the use of antibiotics. This information will be useful in assessing how changes in culturable microbiota may affect the health of farmed European seabass, indicating a potential problem for fish health management during disease outbreaks.

Keywords: bacteria; European seabass aquaculture; Adriatic Sea; antibiotic resistance; *Pseudomonas*; *Vibrio*

Citation: Ramljak, A.; Vardić Smrzlić, I.; Kapetanović, D.; Barac, F.; Kolda, A.; Perić, L.; Balenović, I.; Klanjšček, T.; Gavrilović, A. Skin Culturable Microbiota in Farmed European Seabass (*Dicentrarchus labrax*) in Two Aquacultures with and without Antibiotic Use. *J. Mar. Sci. Eng.* **2022**, *10*, 303. https://doi.org/10.3390/jmse10030303

Academic Editor: Snježana Zrnčić

Received: 29 December 2021
Accepted: 11 February 2022
Published: 22 February 2022

Publisher's Note: MDPI stays neutral with regard to jurisdictional claims in published maps and institutional affiliations.

1. Introduction

Common fish diseases and environmental pollution can lead to economic losses and poor and unprofitable production in aquaculture [1]. European seabass (*Dicentrarchus labrax*) is one of the most farmed fish species in Croatia and tops the list of most farmed fish species in the European Union [2]. Its cultivation is threatened by several common bacterial pathogens and the infections that are caused by them. For example, *Photobacterium damselae* causes photobacteriosis in European seabass with a mortality rate between 60% and 80% [1]. *Vibrio* species are ubiquitous in aquatic ecosystems, while many *Vibrio* species are serious opportunistic pathogens causing the most common bacterial diseases [3–5]. Control of bacterial pathogens in aquaculture production is routinely achieved by the administration of antibiotics. However, excessive antibiotic use has led to the emergence of antibiotic-resistant bacteria [1].

As the skin, along with the gills, is an organ that is involved in the primary defense of the organism from pathogens, it is important to investigate its microflora and how microflora disorders lead to disease development. Microflora refers to the microorganisms that are present on the mucous membranes and skin of fish [6]. Although microorganisms that are pathogenic to fish are part of the normal fish microflora, they do not cause disease unless the balance of the microflora is disturbed Disturbances in the balance (dysbiosis) are often caused by the specificities of aquaculture such as high stocking density and low oxygen concentration [7], but also by changes of physical and chemical conditions in water, temperature fluctuations, seasonality, climate variability, and the use of antibiotics [8]. The skin microbiota of the fish that are exposed to such stress is usually composed of microorganisms that are pathogenic to fish [7]. The viscoelasticity of the skin mucus blocks various bacteria, and upon contact with the bacterial pathogen, the fish excrete more mucus and alter the composition of the skin microflora [9].

Most of the research on European seabass microflora has focused on the gut microbiota that is exposed to conventional [10] and alternative diets [11], as well as on the effects of probiotics on the balance of the gut microbiota [12]. Several researchers focus on the seasonal survey of the marine aquaculture microbiome in a European seabass farm [13], the assessment of seawater microbial quality, and the health status of farmed European seabass [14], with more specific characterization of isolated *Vibrio* species [5]. The evaluation of the tissue-specific diversity of microbiomes within and between sea bass and sea bream [7], as well as the recent study of the effects of aging on the skin and gill microbiota of farmed European sea bass and sea bream, were done by Rosado et al. [7,15].

Although studies of the European seabass skin microbiota are not numerous [3,6,7], one study examined the basic diversity of skin microbiomes and gills of farmed seabass and gilthead sea bream [7], but the link to the history of antibiotic therapy remained unclear. Some recent reports have studied the health problems of European seabass in aquaculture [3,6], but only one correlates the abundance of bacteria with problems that are associated with subsequent antibiotic treatment with oxytetracycline [6]. The abundance of taxa belonging to the non-pathogenic marine group NS3a, and *Polaribacter* 4 decreased in the skin microbiome of diseased fish with oxytetracycline therapy [6], while the pathogens *Pseudomonas* and *Stenotrophomonas* increased significantly. However, there is a lack of available information on the microbiota in the skin of European seabass that are grown in farming practices without a history of antibiotic therapy. Due to the increasing importance of aquaculture, more research is needed in disease prevention and alternative treatments to prevent bacterial resistance to antibiotics [1].

In the present study, we characterized the culturable bacteria of the skin of healthy adult European seabass during summer commercial harvest in two cage fish farms. This included one conventional farm where antibiotics are used to treat disease, and the other practicing a certified antibiotic-free production. The main objectives of the research are: to compare the culturable bacteria of farmed European seabass between these two fish farms and to identify the potential opportunistic bacterial pathogens and determine their susceptibility to antibiotics.

2. Materials and Methods

2.1. Sampling

This study was conducted on two floating cage fish farms, NAS farm in the northern, and MAS farm in the middle Adriatic (Figure 1). Both farms are in the semi open sea, with the NAS farm at depth of approximately 49 m, and MAS farm depth of approximately 22 m. The cages that are 10 m deep contain European seabass (*Dicentrarchus labrax*) and sea bream (*Sparus aurata*). Samples of fish, seawater, and sediments were collected during the summer at both farms. During the sampling year, the NAS farm was subjected to regular periodical control to determine antibiotic residues by an independent certification control service. Antibiotic-free production has been practiced since the beginning of aquaculture production on the NAS farm. The MAS farm differed from the NAS farm with respect to the usage of antibiotics. The antibiotics were administered at therapeutic dosage by a professional veterinarian at the farm based on a positive diagnosis. Usually, antibiotic treatment is conducted for the full period that is required for therapy and as little as necessary. The available information concerning antibiotic administration at the NAS farm have been undisclosed, however it is known that most used antibiotics in the Mediterranean countries are flumequine and oxytetracycline [16]. No clinical signs of fish disease were observed during sampling on both farms, and no antibiotic was used on the MAS farm at the time of sampling. The fish samples were obtained from cages of European seabass that were harvested for commercial purposes at the NAS and MAS farm. A total of 10 individual fish were obtained during commercial catch at each farm. One skin swab per fish was taken from 1 cm^2 areas below the dorsal fin using sterile rods with a 1 cm cotton tip (Deltalab, Barcelona, Spain) for isolation and later bacterial identification. We specifically chose this area for swab collection because this area was not affected by fish handling during sampling which could have caused unwanted contamination. The samples were then serially diluted in 10 mL of sterile phosphate buffer saline (PBS, Sigma, St. Louis, MO, USA) [17]. During sampling, the health status of the fish was assessed and biometric data, namely, the total length and total weight, were recorded.

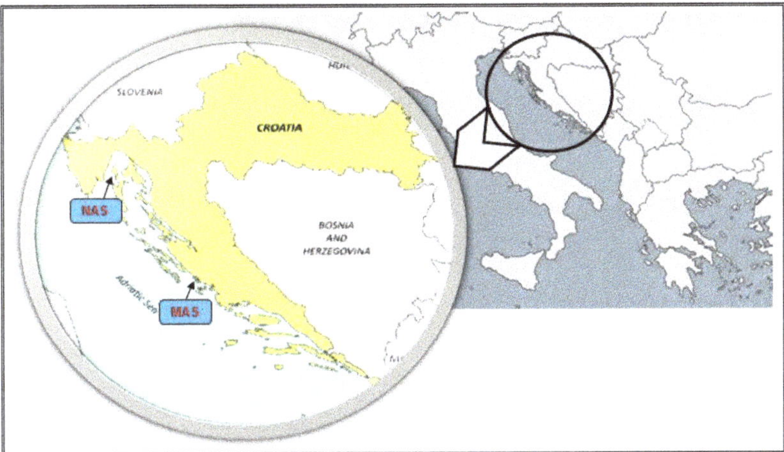

Figure 1. Sampling locations: NAS—marine fish farm in the north Adriatic and MAS—marine fish farm in the middle Adriatic.

Sediment and seawater samples were collected from the NAS and MAS farm below the cages containing the European sea bass. Sediment samples (10 g of the top sediment layer) were collected from each farm using Ekman grab. Seawater samples were collected at four different depths (0.5 m below the surface, 6 m deep, 12 m deep, and 0.5 m above the bottom) using a Niskin water sampler and poured in sterile 0.5 L bottles. All the samples were serially diluted in 10 mL of sterile PBS solution (Sigma) and counts were determined by inoculating the undiluted and serially diluted samples on appropriate media for enumeration and isolation for microbial and molecular analysis (see below).

2.2. Physicochemical and Microbiological Analysis of Seawater and Sediment

The measurements of physicochemical parameters of seawater were carried out at both farms. The pH was measured electrometrically with portable digital SevenGo pro/Ion multiparameter probes (Mettler Toledo, Schwerzenbach, Switzerland) with an accuracy of 0.001 mg/L. The concentration of dissolved oxygen (mg/L), oxygen saturation of water (%), and the temperature (°C) were measured with a SevenGo pro/SG9 OptiOx probe (Mettler Toledo) with a precision of 0.01 mg/L. Conductivity and total dissolved solids (TDS) were measured with a SevenGo pro/conductivity probe (Mettler Toledo) with a precision of 0.1 mg/L.

The total coliforms and *Escherichia coli* were determined using Colilert-18™ and Quanti-Tray/2000 substrate technology (IDEXX, Westbrook, ME, USA). After incubation for 24 h at 35 °C, the appearance of a yellow color of the chamber allowed the evaluation of total coliforms, and fluorescence under UV light indicates the presence of *E. coli*. Enterococci were determined using Enterolert-E™ and Quanti-Tray/2000 (IDEXX). After incubation for 24 h at 41 °C, the appearance of fluorescence indicates the presence of enterococci [18]. Quanti-Tray/2000 indicates the most probable number of bacteria (MPN) in a 100 mL sample using the manufacturer's reagents.

2.3. Number of Heterotrophic Bacteria and Vibrio Count

The number of heterotrophic bacteria (Heterotrophic Plate Count (HPC)) of the skin, seawater, and sediment, was determined by the spread plate method on Difco™ Marine Agar 2216 medium (BD, Sparks, MD, USA) that was incubated at 22 °C for 3–5 days. The spread plate method on a selective Thiosulfate-Citrate-Bile-Sucrose (TCBS) medium (BD), was used for the isolation of the bacteria of the genus *Vibrio* from fish skin swabs, sediment, and seawater samples. The plates were incubated at 22 °C for 24 h. The results are reported as the mean number of colony forming units (CFU) in 1 mL of sediment and seawater or per 1 cm^2 of skin $\pm$ the standard deviation of two technical replicates by sample type [18]. Subsequently, two bacterial colonies representing different morphologies per plate on Marine Agar and TCBS agar were selected from each sample and transferred to Difco™ Tryptic Soy Agar (TSA) (BD) with the addition of 1% NaCl (Kemika, Zagreb, Croatia) (MTSA) plates to obtain a pure culture. After purification and plating on the MTSA plates, a total of 146 bacterial isolates were obtained.

2.4. DNA Isolation and PCR Amplification of Partial 16S rRNA Gene

A small amount of every purified bacterial colony was taken by a sterile loop and subjected to DNA isolation using GenEluteTM Mammalian Genomic DNA 132 Miniprep Kit (Sigma) according to the manufacturer's instructions. Partial 16S rRNA gene sequence was amplified in PCR reaction mixtures containing 1× EmeraldAmp® GT PCR Master Mix (Takara, Shiga, Japan), 0.4 pmol/µL of primers 27F and 1492R (Wilson et al., 1990), 2 µL of DNA template, and nuclease-free water to the final volume of 50 µL. The reaction conditions were described previously [18]. Electrophoresis in 1.5% agarose gel was performed to check the presence of products of approximately 1450 bp length.

2.5. Sequencing and Phylogenetic Analysis of Vibrio sp.

The amplified PCR products were further sequenced commercially by Macrogen (Amsterdam, The Netherlands). The obtained sequences were edited manually and/or in BioEdit version 7.2.5. [19] and deposited in the GenBank under accession numbers: OL979296—OL979441. The sequences were analyzed by comparison with previously characterized 16S rRNA gene of the closest bacteria from the GenBank database using the NCBI BLAST program (Bethesda, MD, USA) and the percentage of similarity is highlighted.

To differentiate *Vibrio* species phylogenetic analysis that was based on the 16S rRNA gene sequence was conducted using MEGA11 [20]. The evolutionary history was inferred by using the maximum likelihood method and Kimura 2-parameter model. A discrete Gamma distribution was used to model the evolutionary rate differences among sites (5 categories (+G, parameter = 0.3935). The tree is drawn to scale, with branch lengths measured in the number of substitutions per site. This analysis involved 69 nucleotide sequences, 15 of which were references from GenBank (accession numbers shown in the Table S1 in Supplementary File). There was a total of 1724 positions per sequence in the final dataset.

2.6. Antimicrobial Susceptibility

The Kirby–Bauer disk diffusion method on BBL™ Mueller Hinton II agar (BD) was used to determine the antimicrobial susceptibility of the isolated bacterial strains. The most used antibiotics in aquaculture were selected. The following antimicrobial disks were used for the test (the amounts are given in micrograms in parentheses): ampicillin (10), streptomycin (10), gentamicin (10), chloramphenicol (30), ciprofloxacin (5), erythromycin (15), imipenem (10), oxytetracycline (30), sulfamethoxazole/trimethoprim (23.75/1.25), vancomycin (30) that is manufactured by BBL™ Sensi-Disk™, and enrofloxacin (5), florfenicol (30), and flumequine (30) that is manufactured by Thermo Scientific™ Oxoid (Hampshire, UK). The inoculum was prepared in 5 mL of sterile 0.85% suspension medium (BioMérieux). On the Vitek Systems ATB 1550 instrument (BioMérieux, Marcy l'Etoile, France), the turbidity for each inoculum was adjusted to 0.5 according to the McFarland value scale [14]. After 24 h incubation at 22 °C, the diameter of the zone of inhibition was measured with a ruler and the values were interpreted as sensitive, moderately sensitive, or resistant according to the manufacturer's instructions.

2.7. Statistical Analysis

Statistical analysis was applied to detect the differences of heterotrophic bacteria and *Vibrio* content between the skin swabs, the seawater, and the sediment samples from both farms. Statistical significance of differences in the number of antibiotic-resistant bacterial isolates between the farms in each type of sample (skin, seawater, sediment) was also analyzed. The non-parametric Mann–Whitney U-test was applied for statistical analyses using SigmaPlot version 14.0 (Systat Software Inc., San Jose, CA, USA). The observed differences were statistically significant at $p < 0.05$. Venn diagrams were drawn to visualize the NAS and MAS culturable bacteria using a freely available web tool (http://bioinformatics.psb.ugent.be/webtools/Venn/, accessed on 15 December 2021).

3. Results

3.1. Results of Sea Bass Health Examination

An examination of the health status of the European seabass from both cage farms revealed no clinical signs of disease. In the NAS farm, the total length of the fish ranged from 28.5 cm to 34.8 cm and the total weight ranged from 281.5 g to 487.7 g. In the MAS farm, the total length of the sampled European seabass ranged from 30.9 cm to 43.2 cm, and the total weight ranged from 254.2 g to 686.5 g. The total body length was statistically significantly higher for the fish from farm MAS ($p < 0.05$) (Figure 2).

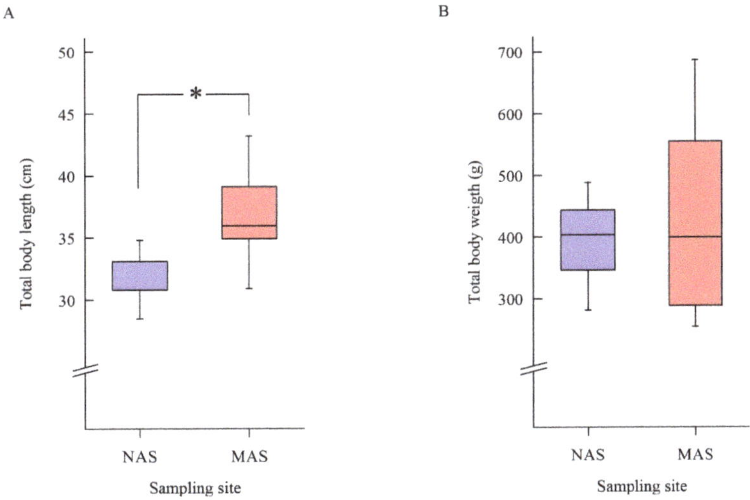

Figure 2. (**A**) The total body length (cm) and (**B**) the total body weight (g) of sea bass (N = 10) from sites NAS (■) and MAS (■). Square boxes indicate the lower and upper quartiles and the whiskers represent the minimum and maximum data values (1.5 interquartile range). The medians are depicted by a solid line. * Significant difference ($p < 0.05$) between the sites.

3.2. Physicochemical and Microbiological Analysis of Seawater

Table 1 shows the physicochemical parameters of the seawater from both fish farms.

Table 1. Physicochemical analysis of seawater from the NAS and MAS fish farms.

Depth (m)	Secchi (m)	Sal. (ppt)	Cond. (μS/cm)	TDS (mg/L)	Temp. (°C)	pH	DO (mg/L)	DO (%)
NAS	28 m							
0.5 m		38.25	51.5	37.3	22.7	8.35	7.01	96.4
6 m		38.26	51.5	37.3	21.7	8.36	6.94	95.4
12 m		38.27	51.5	37.3	19.7	8.37	6.93	95.3
0.5 m above the bottom		38.43	46.8	37.5	15.1	8.23	6.62	83.7
MAS	20 m							
0.5 m		37.3	56.2	28.1	23.5	8.12	8.48	99.2
6 m		37.1	56.0	28.0	23.0	8.13	8.07	93.4
12 m		37.2	56.3	28.0	22.5	8.13	8.35	95.5
0.5 m above the bottom		37.2	56.2	28.1	20.1	8.16	8.02	87.6

NAS—North Adriatic Sea antibiotic-free farm; MAS—Middle Adriatic Sea conventional farm; Secchi—Secchi depth; Sal.—Salinity; Cond.—Conductivity; TDS—Total dissolved solids; Temp.—Temperature; DO (mg/L)—Dissolved oxygen values; DO%—Dissolved oxygen saturation.

The results of the microbiological analysis of the seawater and sediment from the two farms are presented in Table 2. Although the levels of total coliform bacteria in the seawater column were notably higher at the NAS farm, the difference between two farms was not statistically significant. The levels of *E. coli* were higher at the MAS farm, but also with no statistically significant difference ($p > 0.05$).

Table 2. Microbiological analysis of seawater and sediment from the NAS and MAS farms.

	NAS			MAS		
Sample Type	TC (MPN/100 mL)	EC (MPN/100 mL)	EN (MPN/100 mL)	TC (MPN/100 mL)	EC (MPN/100 mL)	EN (MPN/100 mL)
Seawater						
0.5 m	487.0	<10.0	<10.0	88.0	25.5	10.0
6 m	588.0	<10.0	<10.0	81.5	10.0	< 10.0
12 m	1034.0	<10.0	<10.0	20.0	10.0	< 10.0
0.5 m above the bottom	10.0	<10.0	<10.0	20.5	< 10.0	< 10.0
Sediment	<10.0	<10.0	<10.0	15.0	< 10.0	46.5

NAS—North Adriatic Sea antibiotics-free farm; MAS—Middle Adriatic Sea conventional farm; TC—Total coliforms; EC—*E. coli*; EN—Enterococci.

The studied parameters in the sediment samples had higher values in the samples from the MAS farm, except for the number of *E. coli*, which had the same value in the sediment samples from both farms.

3.3. Number of Heterotrophic Bacteria and Vibrio Count

The results of the microbiological analysis of skin swabs of European seabass from the NAS and the MAS farm are presented in Table 3. The total number of heterotrophic bacteria (HPC) from the skin swabs of European seabass between the farms was statistically significantly higher in the NAS antibiotics-free farm, as well as the number of bacteria of the genus *Vibrio* ($p < 0.05$).

Table 3. Microbiological analysis of heterotrophic bacteria (HPC) and *Vibrio* count from skin swabs of European seabass, seawater, and sediment from NAS and MAS farm.

	NAS			MAS		
	Skin	Water	Sediment	Skin	Water	Sediment
HPC (CFU/mL)	69.1 ± 21.7	278.75 ± 196.3	60 ± 14.1	36.2 ± 20.0	480 ± 281.9	120 ± 11.3
Vibrio (CFU/mL)	38.1 ± 35.2	14 ± 3.6	3 ± 0	5 ± 7.1	55 ± 51.1	88 ± 4.2

The differences in the total number of heterotrophic bacteria in seawater between the farms were not statistically significant ($p > 0.05$). The number of bacteria of the genus *Vibrio* was higher in the seawater samples from the MAS farm, and the differences were statistically significantly higher than in the NAS farm ($p < 0.05$).

3.4. Culturable Microbiota

The 16S rRNA gene sequences were generated for 81 and 64 bacterial isolates from NAS and MAS farm, respectively. The results from the datasets of each fish farm were composed of three compartments: the skin swabs of European seabass, seawater, and sediments. Of the total number of bacterial isolates from the NAS farm, 28 were from skin swabs, 32 were from seawater, and 21 were from sediments. Based on the sequence comparison, 33 bacterial species were identified among the isolates from the NAS farm, which are listed in Table 4. Of them, two, namely *V. alginolyticus* and *V. harveyi*, were identified in both the skin and seawater. Only *Bacillus hwajinpoensis* was found in the skin and sediment (Figure 3).

Table 4. Results of molecular identification of bacterial isolates from the NAS antibiotic-free farm.

Sample Type		Skin (*n* = 28)		Seawater (*n* = 32)		Sediment (*n* = 21)	
Species	Percent Identity %	No.	%	No.	%	No.	%
Alcaligenes faecalis	97.9	1	3.6				
Aliivibrio finisterrensis	99.0			1	3.1		
Alteromonas macleodii	99.6			2	6.3		
Bacillus aquimaris	99.7–99.8					7	33.3
Bacillus horikoshii	99.6					1	4.8
Bacillus hwajinpoensis	99.4–100.0	2	7.1			3	14.3
Bacillus idriensis	99.8					1	4.8
Bacillus tianshenii	99.7					1	4.8
Microbacterium oxydans	99.6	1	3.6				
Paenisporosarcina quisquiliarum	99.5					1	4.8
Photobacterium aphoticum	99.0–99.9					4	19.0
Pseudoalteromonas arabiensis	99.2–99.5	3	10.7				
Pseudoalteromonas hodoensis	99.5			1	3.1		
Pseudoalteromonas phenolica	99.8			1	3.1		
Pseudoalteromonas shioyasakiensis	99.4–99.7			3	9.4		
Pseudoalteromonas tetraodonis	99.7–99.9			3	9.4		
Pseudoalteromonas undina	100.0	1	3.6				
Pseudochrobactrum saccharolyticum	98.9–99.5	4	14.3				
Pseudomonas zhaodongensis	99.4	1	3.6				
Shewanella marinintestina	99.9					1	4.8
Vibrio alginolyticus	99.6–100.0	5	17.9	2	6.3		
Vibrio chagasii	98.2–98.8			6	18.8		
Vibrio crassostreae	99.1			1	3.1		
Vibrio cyclitrophicus	99.4–100.0	4	14.3				
Vibrio europaeus	99.7			1	3.1		
Vibrio fortis	99.1–99.3			2	6.3		
Vibrio gigantis	99.6			1	3.1		
Vibrio harveyi	99.7–100.0	4	14.3	1	3.1		
Vibrio hyugaensis	99.7			1	3.1		
Vibrio kanaloae	99.2–100.0			3	9.4		
Vibrio neocaledonicus	99.8	2	7.1				
Vibrio toranzoniae	98.5–99.9					2	9.5
Vibrio tubiashii	96.1–99.3			3	9.4		

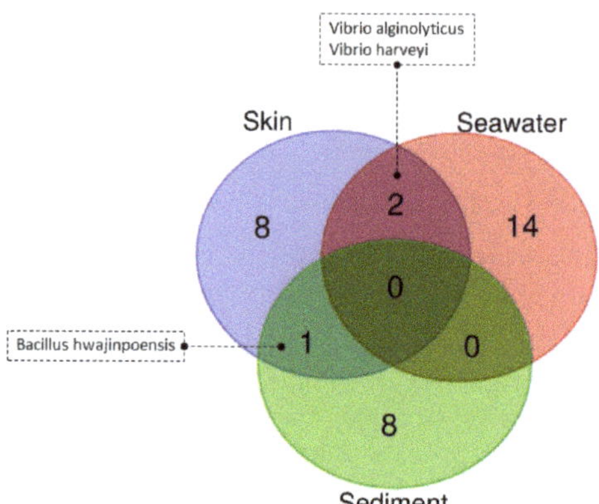

Figure 3. Venn diagram of the bacterial species that were identified in culturable microbiota of fish skin, seawater, and sediment associated with the NAS marine fish farm.

At the MAS farm, 20 isolates out of 64 in total were isolated from the European seabass swabs, while 22 accounted for seawater and sediment isolates. A total of 22 bacterial species were identified among the isolates from all the sample types in the MAS farm (Table 5). Despite the difference of species number (Table 5) among the bacterial communities of the fish skin (11 species), seawater (3 species), and sediment (4 species), species *Marinobacter litoralis*, and two species from the genus *Pseudoalteromonas* (*Pseudoalteromonas tetraodonis, Pseudoalteromonas undina*) were detected in the seawater and sediment, whereas *V. toranozoniae* was identified in the skin and sediment (Figure 4).

Table 5. Results of molecular identification of the bacterial isolates from the MAS conventional farm.

Sample Type		Skin (*n* = 20)		Seawater (*n* = 22)		Sediment (*n* = 22)	
Species	Percent Identity %	No.	%	No.	%	No.	%
Achromobacter spanius	98.7	1	5.0				
Aeromonas molluscorum	99.7	1	5.0				
Agrococcus sp.	98.9	1	5.0				
Erwinia billingiae	99.6	1	5.0				
Ewingella americana	99.3–99.4	2	10.0				
Halomonas aquamarina	99.5			1	4.5		
Halomonas boliviensis	99.8					1	4.5
Marinobacter litoralis	99.4–100.0			13	59.1	2	9.1
Paenalcaligenes suwonensis	99.6–99.7					2	9.1
Photobacterium lutimaris	98.4–98.8					4	18.2
Pseudoalteromonas tetraodonis	99.9			2	9.1	4	18.2
Pseudoalteromonas undina	99.5–99.8			4	18.2	1	4.5
Pseudomonas azotoformans	99.8	1	5.0				
Pseudomonas gessardii	99.8	1	5.0				
Pseudomonas kribbensis	99.9			1	4.5		
Pseudomonas poae	98.8	1	5.0				
Pseudomonas sp. DSM 28142	99.9	1	5.0				
Pseudomonas zhaodongensis	99.7			1	4.5		
Shewanella arctica	99.0–99.5	3	15.0				
Vibrio anguillarum	98.8–99.7	4	20.0				
Vibrio kanaloae	96.3–99.5					2	9.1
Vibrio toranzoniae	99.4–100.0	3	15.0			6	27.3

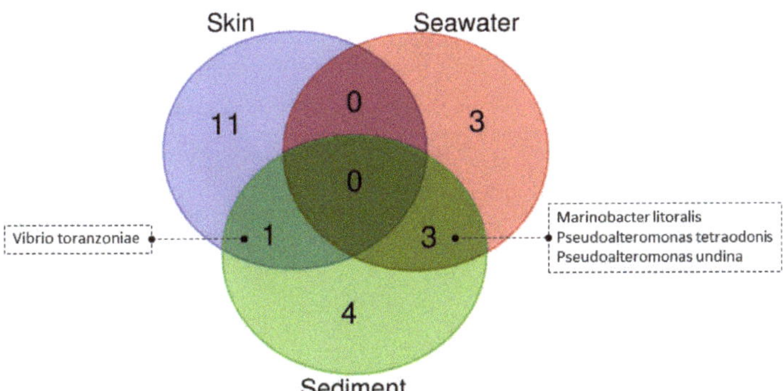

Figure 4. Venn diagram of the bacterial species that were identified in culturable microbiota of fish skin, seawater, and sediment associated with marine fish farm MAS.

A total of seven bacterial genera were identified by molecular analysis of bacterial isolates (N = 28) from skin swabs of European seabass from the NAS antibiotic-free farm. Figure 5A shows the occurrence of bacterial genera of skin swabs. The most common bacteria were bacteria of the genus *Vibrio* (53%), followed by the genera *Pseudoalteromonas* (14%) and *Pseudochrobactrum* (14%). *V. alginolyticus* (17.9%), and *Pseudochrobactrum sacchaloryticum* (14.3%), where *V. cyclitrophicus* (14.3%) and *V. harveyi* (14.3%) accounted for the largest number of the 11 bacterial species that were isolated (Table 4).

At the NAS antibiotics-free farm, the greatest diversity of bacterial genera (seven genera) was found in samples of skin swabs of European seabass. A total of five bacterial genera were detected in the sediment samples, while four bacterial genera were detected in the seawater samples. The most dominant bacterial genus that was identified in the seawater samples was *Vibrio* (65%), followed by *Pseudoalteromonas* (25%) (Figure 5C). Among the bacterial species, *V. chagasii* (18.8%), *Pseudoalteromonas shioyasakiensis* (9.4%), *Pseudoalteromonas tetraodonis* (9.4%), *V. kanaloae* (9.4%), and *V. tubiashii* (9.4%) were the most frequently identified.

In the sediment samples from the NAS farm, 21 bacterial isolates were obtained from five bacterial genera, among which the genus *Bacillus* (62%) dominated (Figure 5E). *B. aquimaris* (33.3%), *Photobacterium aphoticum* (19.0%), and *B. hwajinpoensis* (14.3%) corresponded to the most frequently identified bacterial species (Table 4).

V. alginolyticus and *V. harveyi* were isolated from European seabass skin swabs and seawater at the NAS antibiotic-free farm. *B. hwajinpoensis* was isolated from the skin swabs and sediment at the same farm. The seawater samples displayed the highest diversity of bacterial species (N = 16). A total of 11 bacterial species were identified from the skin swabs from the European seabass samples, while nine species were identified in the sediment samples (Table 4).

At the MAS conventional farm, among 20 bacterial isolates from skin swabs of European seabass, eight bacterial genera were identified, as shown in Figure 5B. The dominant bacterial genera were *Vibrio* (35%) and *Pseudomonas* (20%). Of the 12 bacterial species from the skin swabs of European seabass, *V. anguillarum* (20%), *Shewanella arctica* (15%), and *V. toranzoniae* (15%) were the most abundant (Table 5).

A total of four bacterial genera were identified in seawater samples from the MAS farm (22 bacterial isolates), of which the genus *Marinobacter* (59%) and the genus *Pseudoalteromonas* (27%) were the most abundant. The results of the presence of bacterial genera in the seawater samples are shown in Figure 5D. *M. litoralis* was most frequently identified bacterial species (59.1% of the total seawater isolates) (Table 5).

The identified bacterial genera from the sediment of the MAS farm and their representation are shown in Figure 5F. Among the six bacterial genera, *Vibrio* (36%), *Pseudoalteromonas* (23%), and *Photobacterium* (18%) were the most abundant. The most dominant bacterial species were *V. toranzoniae* (27.3%), *Photobacterium lutimaris* (18.2%), and *Pseudoalteromonas tetraodonis* (18.2%) (Table 5).

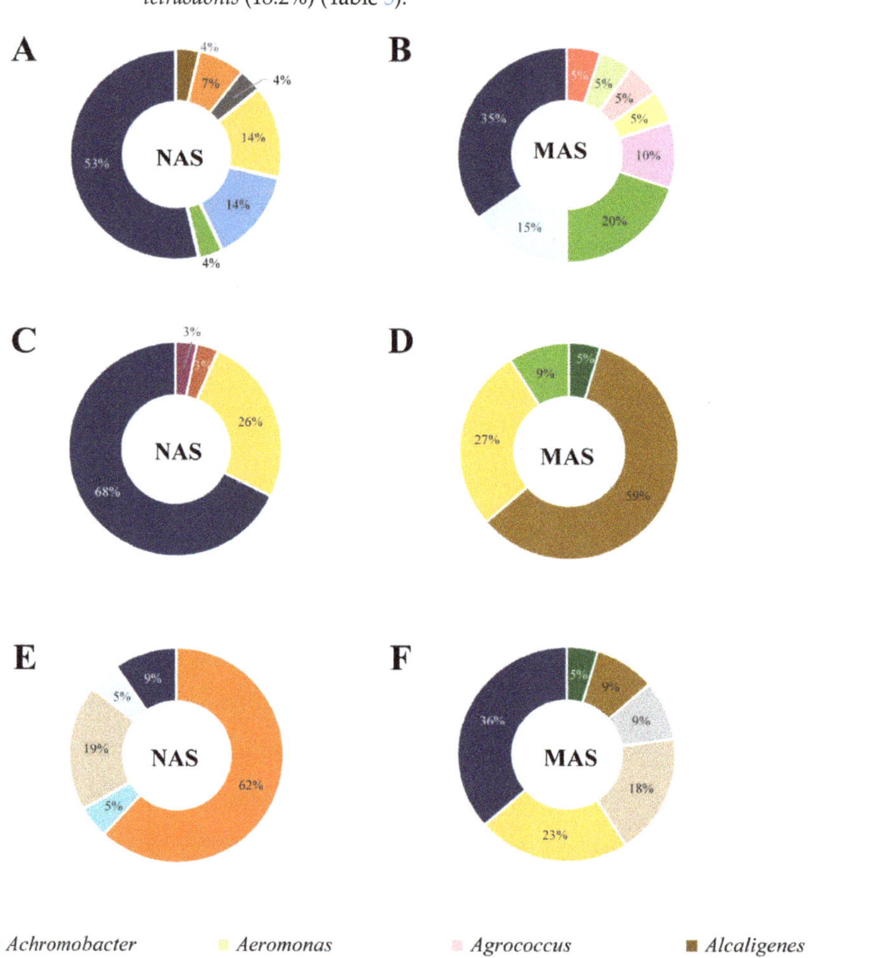

Figure 5. Representation of the bacterial genera (percentage) in: (**A**)—European seabass skin swab samples from the NAS and (**B**)—the MAS farm; (**C**)—in seawater samples from the NAS farm and (**D**)—the MAS farm; (**E**)—in sediment samples from the NAS farm and (**F**)—the MAS farm.

V. toranzoniae was identified in European seabass swab samples and in sediment samples from the MAS farm with antibiotic use. *M. litoralis* was identified in the seawater and in the sediment samples at this farm, as well as *Pseudoalteromonas tetraodonis* and *Pseudoalteromonas undina*. Other bacterial species were detected only in one sample type. Only the genus *Pseudomonas* and the genus *Vibrio* were identified from the swab samples at both farms.

To confirm the identity of *Vibrio* species and check their grouping within different clades, phylogenetic analysis based on 16S rRNA including reference sequences from GenBank was performed. As expected, phylogenetic analysis confirmed grouping into five clades: *V. splendidus*, *V. anguillarum*, *V. orientalis*, *V. fortis*, and *V. harveyi* as well as branching with the same species (Figure 6). In the Tables 4 and 5 the percent of identity with the known *Vibrio* species from GenBank are shown.

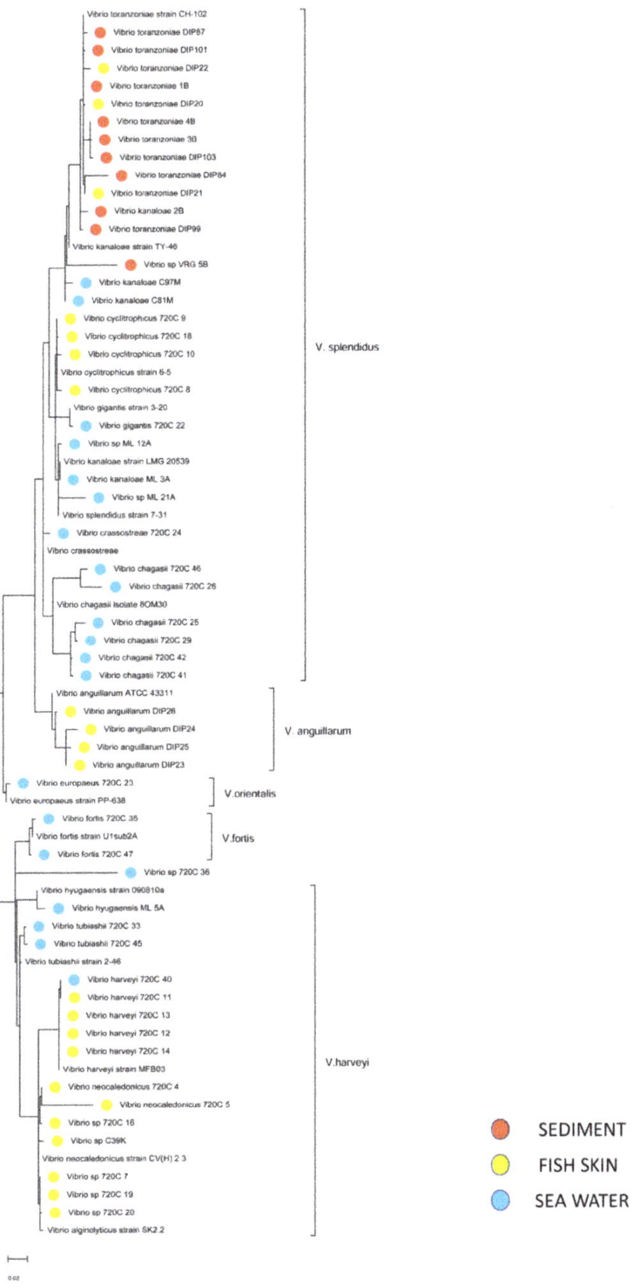

Figure 6. Phylogenetic analysis of studied *Vibrio* species that was inferred from partial 16S rRNA gene sequences using MEGA11 software—the clades are indicated by brackets.

3.5. Antimicrobial Resistance of Bacterial Isolates

The disc diffusion test for susceptibility to 13 antibiotics was performed using the total of 145 bacterial isolates from the two fish farms. The number of isolates that were resistant to individual antibiotics and the percentage of resistant isolates are shown in Table 6.

Table 6. Differences in antibiotic resistance of the bacterial isolates from the NAS antibiotic-free European seabass farm versus farm MAS conventional farm.

| | NAS | | | | | | MAS | | | | | |
| | Skin Swabs (N = 28) | | Seawater (N = 32) | | Sediment (N = 21) | | Skin Swabs (N = 20) | | Seawater (N = 22) | | Sediment (N = 22) | |
Antibiotic	No. of Isolates	%	No. of Isolates	%	No. of Isolates	%	No. of Isolates	%	No. of Isolates	%	No. of Isolates	%
Enrofloxacin	11	39.3	10	31.3	13	61.9	10	50.0	3	13.6	2	9.1
Florfenicol	8	28.6	3	9.4	9	42.9	13	65.0	3	13.6	3	13.6
Gentamicin	4	14.3	6	18.8	4	19.0	4	20.0	1	4.5	-	-
Ampicillin	16	57.1	16	50.0	6	28.6	20	100.0	3	13.6	7	31.8
Erythromycin	11	39.3	12	37.5	9	42.9	15	75.0	3	13.6	6	27.3
Oxytetracycline	9	32.1	1	3.1	6	28.6	7	35.0	3	13.6	2	9.1
Sulfamethoxazole/Trimethoprim	6	21.4	3	9.4	3	14.3	11	55.0	6	27.3	2	9.1
Vancomycin	24	85.7	27	84.4	9	42.9	20	100.0	20	90.9	19	86.4
Flumequine	6	21.4	4	12.5	11	52.4	8	40.0	4	18.2	4	18.2
Imipenem	5	17.9	2	6.3	6	28.6	13	65.0	1	4.5	2	9.1
Ciprofloxacin	9	32.1	6	18.8	10	47.6	6	30.0	4	18.2	4	18.2
Streptomycin	11	39.3	13	40.6	6	28.6	11	55.0	3	13.6	2	9.1
Chloramphenicol	4	14.3	1	3.1	3	14.3	8	40.0	2	9.1	1	4.5

N—number of analyzed bacterial isolates, No. of isolates—number of resistant bacterial isolates, %—percentage of resistant bacterial isolates.

Most of the bacterial isolates from the skin swabs of European seabass from the NAS farm without antibiotic use showed resistance to vancomycin (85.7%), ampicillin (57.1%), enrofloxacin (39.3%), erythromycin (39.3%), and streptomycin (39.3%). All the isolates of *V. alginolyticus* were resistant to vancomycin and ampicillin (Table 7). The resistance of isolates to other antibiotics was inconsistent.

Table 7. Antibiotic resistance of bacteria from the genus *Vibrio*.

| | *V. alginolyticus* (N = 7) | | *V. toranzoniae* (N = 11) | | *V. anguillarum* (N = 4) | | *V. chagasii* (N = 6) | | *V. harveyi* (N = 5) | |
Antibiotic	No. of Isolates	%	No. of Isolates	%	No. of Isolates	%	No. of Isolates	%	No. of Isolates	%
Enrofloxacin	0		2	18	4	100	2	40	2	40
Florfenicol	0		0		3	75	0		2	40
Gentamicin	0		0		2	50	0		0	
Ampicillin	7	100	4	36	4	100	6	100	5	100
Erythromycin	1	14	2	18	4	100	4	67	5	100
Oxytetracycline	1	14	0		1	25	0		0	
Sulfamethoxazole/Trimethoprim	0		1	9	1	25	0		0	
Vancomycin	7	100	8	73	4	100	6	100	5	100
Flumequine	0		3	27	3	75	1	17	1	20
Imipenem	1	14	4	36	3	75	0		1	20
Ciprofloxacin	1	14	2	18	2	50	1	17	3	60
Streptomycin	1	14	1	9	4	100	5	83	2	
Chloramphenicol	0		2	18	1	25	0		0	

N—number of analyzed bacterial isolates, No. of isolates—number of resistant bacte-rial isolates, %—percentage of resistant bacterial isolates.

Most of the bacterial isolates from the skin swabs of European seabass from the MAS conventional farm with antibiotic use were resistant to ampicillin (100.0%), vancomycin (100.0%), erythromycin (75.0%), florfenicol (65.0%), and imipenem (65.0%). The predominant bacterial species in the samples was *V. anguillarum* and the isolates were resistant to enrofloxacin, ampicillin, erythromycin, vancomycin, and streptomycin (Table 7). No statistically significant differences in the number of antibiotic-resistant bacterial isolates were found between the two farms ($p > 0.05$).

Similar to the isolates from the skin swabs, most bacterial isolates from seawater from the NAS antibiotics-free farm showed resistance to vancomycin (84.4%), ampicillin (50.0%), enrofloxacin (31.3%), erythromycin (37.5%), and streptomycin (40.6%). The predominant bacterial species in the seawater samples was *V. chagasii* and all *V. chagasii* isolates showed resistance to vancomycin (Table 7).

Among the bacterial isolates from the farm seawater samples at the MAS farm, vancomycin resistance was the most common (90.9%). In seawater samples, the most common bacterial species was *M. litoralis* and all *M. litoralis* isolates were resistant to vancomycin, while resistance was not consistent among the other antibiotics. The differences in the number of resistant bacterial isolates between the two farms were not statistically significant ($p > 0.05$).

Most of the bacterial isolates from the sediment samples from the NAS antibiotics-free farm showed resistance to enrofloxacin (61.9%) and flumequine (52.9%). *B. aquimaris* was the most common isolate and most of the bacterial isolates (six out of seven) were resistant to enrofloxacin. For other antibiotics, the resistance of the bacterial isolates was not uniform.

As high as 86.4% of total bacterial isolates from the MAS farm sediment samples were resistant to vancomycin. This is also characteristic for the most numerous sediment isolates that were identified as *V. toranzoniae* (Table 7). In contrast to the results of resistance from the skin swabs and seawater isolates, the number of antibiotic-resistant bacterial isolates between the two farms were statistically significantly higher at the NAS antibiotic-free farm than in the MAS conventional farm ($p < 0.05$).

4. Discussion

The quality of seawater has a great influence on the success of fish farming. Stress that is caused by disturbances in seawater quality negatively affects fish growth and development [21]. The seawater physicochemical parameters that were measured at the NAS and MAS farms were consistent with previous studies on Adriatic Sea bass farms [17,22,23].

Microbiological analysis of the seawater samples and skin swabs of European seabass revealed a higher HPC and *Vibrio* count in the seawater samples from the MAS farm, but conversely higher HPC and *Vibrio* count from skin swabs from the NAS farm. This HPC count from the skin swabs of European seabass from the NAS farm (69.1 ± 21.7 CFU/cm^2) is lower than a previously reported range between 10^2 and 10^4 CFU/cm^2, at other geographical locations [8]. The higher *Vibrio* count in the seawater samples from the MAS farm can be explained by the influence of the higher seawater temperature which stimulates bacterial growth [22,24]. Interestingly, a higher *Vibrio* count was detected in samples from the fish skin than in the seawater of the NAS farm. This phenomenon was described by Vatsos [25] who documented the quantitative and qualitative differences between the culturable microbiota of the fish skin and that in the water from the host environment. Although the identification of *Vibrio* species is challenging due to similarities, both in phenotypic properties and 16S rRNA sequences, phylogenetic analyses can differentiate clades and subgroups affiliation [26]. That is why we used this analysis for the *Vibrio* species differentiation.

In this study, a similar number of bacterial genera was observed in each type of sample (skin swabs, seawater, sediment) after analyzing the composition of the culturable microbiota from the two investigated farms (under culture conditions with and without the use of antibiotics). This could be because each bacterial species grows under specific culture conditions. A total of four genera were found in the seawater samples from both farms, but with different prevalence. The number of bacterial species differed between the seawater samples from the NAS antibiotics-free farm and the MAS conventional farm, where 16 and 6 bacterial species, respectively, were identified. The diversity of bacterial species is strongly influenced by the quality of seawater in the farm [27]. Considering that the physicochemical and microbiological parameters in both farms showed seawater quality that is suitable for farming [17,23], the use of antibiotics at the MAS farm may have affected diversity of the bacterial species. Rosado et al. [6] stated that the use of antibiotics can affect the change in the composition of bacterial species in the skin microbiota, reducing the diversity of bacteria and the resistance of farmed fish to bacterial infections. Nevertheless, in this work, 12 bacterial species were identified from skin swabs in the MAS farm where antibiotics are regularly administered. At the NAS farm, where antibiotics are not used, 11 bacterial species were identified from the skin swabs. However, two common genera, *Pseudomonas* and *Vibrio*, were identified in the skin swab samples at both farms. Our observation agrees with those of Čož Rakovac et al. [28], who previously reported *Pseudomonas* isolates from wild and farmed sea bass in the Northern Adriatic Sea. These results are similar with those of recent study of skin microbiota of farmed European seabass in South Adriatic (Croatia and Montenegro), where most of the isolated bacteria comprised of *Vibrio, Photobacterium,* and *Pseudomonas* genera [16]. The skin is one of the main entry points for pathogens and infections, while the mucus of the fish skin has an important role in the immune system and has an antibacterial effect [1]. Therefore, one of the objectives of this study was to identify bacterial species from skin swabs of European seabass with particular emphasis on pathogenic bacteria to fish. An analysis of the identified bacterial species reveals some potentially pathogenic bacteria to fish [4]. According to the study of De Bruijn et al. [8], the microbiota of fish skin contains pathogenic bacteria to fish that do not cause disease unless the microbiota balance is disturbed. It is well known that antibiotic use can promote the proliferation of opportunistic pathogens [29], e.g., *V. anguillarum* in skin swabs from MAS conventional farm.

Pseudoalteromonas undina was isolated from skin swabs of European seabass from the NAS farm. *P. undina* is widely distributed in seawater, especially in farms that are rich in organic matter, but has not been reported as a pathogenic bacterium, except in a study by Pujalte et al. [30], in which it caused mortalities of sea bass. The opportunistic pathogens *V. alginolyticus* and *V. harveyi* have also been isolated from NAS farms. *V. alginolyticus* is known to cause vibriosis [17,31], while *V. harveyi* is also a potential threat for European seabass farming [5,32].

The samples of European seabass skin swabs from the MAS farm also contained some potentially pathogenic bacteria to fish. *V. anguillarum* is one of the causative agents of vibriosis in the Adriatic Sea [14], while *V. toranzoniae* has only been identified as a pathogen in diseased sea eels (*Genypterus chilensis*) in Chile [33]. In addition, *Pseudomonas gessardii* has been isolated and identified as a larval shrimp opportunistic pathogen in a recent study, while *V. alginolyticus* has been identified as the primary pathogen [31].

Considering the potential pathogenicity of *V. alginolyticus, V. anguillarum,* and *V. harveyi*, it is important to highlight their resistance to antibiotics, although this varied from isolate to isolate. *V. alginolyticus* from the skin swabs of European seabass from a NAS farm showed resistance to streptomycin and ampicillin. In previous studies, *V. alginolyticus* that was isolated from skin swabs of sea bass from three farms in the Adriatic Sea (Lim Bay, Lamjana and Mali Ston Bay) showed higher resistance to five (ampicillin, penicillin, piperacillin, sulfamethoxazole/trimethoprim, and trimethoprim) of the 13 antibiotics that were tested (ampicillin, streptomycin, gentamicin, imipenem, chloramphenicol, florfenicol, ciprofloxacin, enrofloxacin, erythromycin, oxytetracycline, sulfamethoxazole/trimethoprim, vancomycin, and flumequine) [17]. In addition, all the isolates of *V. alginolyticus* in this study were sensitive or moderately sensitive to sulfamethoxazole/trimethoprim, which is commonly used to treat vibriosis; this is consistent with the studies of Zorrilla et al. [34] who tested seven antimicrobial agents (ampicillin, amoxicillin, tetracycline, oxytetracycline, trimethoprim-sulphamethoxazole, oxolinic acid, and flumequine). Isolates of *V. harveyi* from the NAS farm were found to be resistant to ampicillin, erythromycin, and vancomycin. In a similar study, Veić [5] tested nine antibiotics (ampicilin, trimethoprim/sulfadiazine, chloramphenicol, oxytetracycline, enrofloxacin, flumequine, nalidixic acid, gentamicin, and neomycin) and described resistance to gentamicin and neomycin, as well as to ampicillin and erythromycin in all the strains, and intermediate sensitivity to nalidixic acid and oxytetracycline in the same *V. harveyi* strains that were isolated from diseased European seabass in the Adriatic Sea. Kang et al. [35] tested 16 antibiotics (ampicillin, cefotaxime, cefotetan, cephalothin, chloramphenicol, ciprofloxacin, cefepime, erythromycin, gentamicin, kanamycin, nalidixic acid, rifampicin, streptomycin, tetracycline, trimethoprim/sulfamethoxazole, and vancomycin) and reported that *V. harveyi* that was isolated from seawater in South Korea was resistant to ampicillin and vancomycin, and sensitive to erythromycin.

V. anguillarum that was isolated from the MAS conventional farm was resistant to ampicillin, enrofloxacin, erythromycin, streptomycin, and vancomycin. Resistance of *V. anguillarum* that was isolated from farmed rainbow trout (*Oncorhynchus mykiss*) was determined for cloxacillin, ampicillin, sulfamethoxazole/trimethoprim, and erythromycin [36]. Variations in results are commonly attributed to the difference between the tested organism as well as the sampling location. In a recent study that was carried out by Kapetanović et al. [14], *V. anguillarum* that was isolated from the skin and gills of European seabass (Mali Ston Bay) showed resistance to gentamicin, erythromycin, and streptomycin, which is not entirely consistent with the results of this study. Similarly, Veić [5] determined the resistance of isolated *V. anguillarum* strains from diseased European seabass in the Adriatic Sea to ampicillin and erythromycin. Thus, in our study, *V. anguillarum* was found to be more resistant, indicating the possible development of bacterial resistance to certain antibiotics to which resistance was not found in previous studies.

The results of antimicrobial resistance in sediment seem to be somewhat controversial, as resistance is higher in the sediment from the NAS farm where antibiotics are not used. Increased antibacterial resistance of bacteria in sediment is often the most sensitive environmental indicator of past antibiotic use [37]. *Vibrionaceae* were widely distributed in sediment samples on fish farms in the western Mediterranean and their resistance is known due to the antibiotics administered [38]. Interestingly, in our study, antibiotic resistance was detected in the microbiota of the NAS farm where antibiotics were not administered since the aquaculture production was established. Antibiotic resistance depends on the species and the location where the bacteria was isolated. It is well known that antibiotic resistance is higher along the coasts and in sheltered bays than in open waters but could also occur in more remote waters due to the influence of non-aquatic organisms or pollutants [39].

The resistant strains at the antibiotic-free farm could be of terrestrial/agricultural origin. In fact, the nearby island of Cres is known for its traditional and extensive farming of autochthonous sheep, which has previously been diagnosed clinical mastitis, enterotoxaemia, actinobacillosis, contagious ecthyma, foot rot, and *Brucella ovis* infection [40]. The use of various antibiotics for treatment of bacterial diseases in sheep, including those based on enrofloxacin and flumequine [40,41], might promote the development of antibiotic resistance that can potentially be carried towards surface waters and sediments by rainfall runoffs [42]. This should be considered as a potential risk for the release of bacterial pathogens into the water column and the spread of antibiotic resistance across different marine compartments and should be investigated in future studies. There is also the possibility that resistant bacteria were introduced on the NAS farm by fingerlings, which also should be investigated.

These results confirm the presence of some potentially pathogenic bacteria for fish. Bacterial fish diseases are usually treated with antibiotics. The most commonly used antibiotics include tetracyclines and fluoroquinolones. However, there is a risk of antibiotic resistance development, as multidrug-resistant bacteria can be easily spread in the marine environment. In addition, the inadequate treatment of fish diseases with antibiotics can lead to selection of bacteria that respond less to antibiotic treatments and increase the likelihood of disease outbreaks in fish farms. The need to reduce financial losses in European seabass aquaculture, overcome the risk of spreading antibiotic-resistant bacteria, and concerns about environmental impacts and consumer safety require alternative and sustainable control measures for frequent antibiotic therapy in aquaculture.

To our knowledge, this is the first study comparing the culturable microbiota at fish farms with and without the use of antibiotics. Our results should be extended in further studies with a larger number of samplings as well as antibiotic residues analyses in the water and sediment, all of which may affect the results. Before each stocking in the cages, the microbiota of the skin of fish fingerlings should be analyzed and included in further studies. Based on the preliminary results of the present study, further research is needed to confirm these findings and to investigate whether and how the microbiota differs seasonally in these farms with and without antibiotic use.

Supplementary Materials: The following supporting information can be downloaded at: https://www.mdpi.com/article/10.3390/jmse10030303/s1, Table S1: List of bacterial isolates/strains and their accession numbers from GenBank used for the phylogenetic tree construction (Figure 6, Results).

Author Contributions: Conceptualization, D.K., A.R. and A.G.; methodology, A.R., I.V.S., A.K., D.K. and A.G.; validation, I.B. and A.G.; field work, T.K., D.K., A.G. and I.B.; formal analysis, A.R., D.K., I.V.S. and F.B.; investigation, D.K., I.B., F.B. and A.G.; resources, D.K. and I.B.; writing—original draft preparation, A.R., D.K., I.V.S. and L.P.; writing—review and editing, A.K. and A.G.; visualization, A.R., D.K. and I.V.S.; supervision, I.V.S., L.P. and A.G.; project administration, T.K., D.K.; principal investigator, T.K.; funding acquisition, T.K. All authors have read and agreed to the published version of the manuscript.

Funding: This work has been fully supported by the Croatian Science Foundation's funding of the under the project AqADAPT—Grant No IP-2018-01-3150, Croatian Academy of Sciences and Arts, and Croatian Ministry of Science and Education under the bilateral project with Germany (DAAD) EXAQUA.

Institutional Review Board Statement: Not applicable.

Informed Consent Statement: Not applicable.

Data Availability Statement: The data presented in this study are available on request from the corresponding author. The data are not publicly available because it's an ongoing scientific project AqADAPT—Grant No IP-2018-01-31504.

Acknowledgments: This work has been fully supported by the Croatian Science Foundation's funding of the under the project AqADAPT—Grant No IP-2018-01-3150. We also thank the Croatian Academy of Sciences and Arts for funding the project: The presence of resistant bacterial strains as part of the natural microflora of European seabass (*Dicentrarchus labrax*) cultured under conditions with and without the use of antibiotics: truths and prejudices. We thank our colleagues at the Helmholtz Centre for Environmental Research GmbH-UFZ, Leipzig, Germany, who collaborated on the Croatian-German bilateral project: Exploring the potential of phage and microbiomes to control pathogens in aquaculture (EXAQUA). The publication of this article was funded by the Croatian Ministry of Science and Education within the framework of the bilateral project with Germany (DAAD) EXAQUA. We thank the anonymous reviewers for their useful comments and suggestions for our article.

Conflicts of Interest: The authors declare no conflict of interest. The funders had no role in the design of the study; in the collection, analyses, or interpretation of data; in the writing of the manuscript, or in the decision to publish the results.

References

1. Esteban, M.A. An Overview of the Immunological Defenses in Fish Skin. *Int. Sch. Res. Netw.* **2012**, *2012*, 853470. [CrossRef]
2. FAO. 2021. Available online: https://www.fao.org/fishery/countrysector/naso_croatia/en (accessed on 8 November 2021).
3. Camara-Ruiz, M.; Cerezo, I.M.; Guardiola, F.A.; Garcia-Beltran, J.M.; Balebona, M.C.; Morinigo, M.A.; Esteban, M.A. Alteration of the Immune Response and the Microbiota of the Skin during a Natural Infection by Vibrio harveyi in European Seabass (*Dicentrarchus labrax*). *Microorganisms* **2021**, *9*, 964. [CrossRef] [PubMed]
4. Haenen, O.L.M.; Fouz, B.; Amaro, C.; Isern, M.M.; Mikkelsen, H.; Zrncic, S.; Travers, M.A.; Renault, T.; Wardle, R.; Hellström, A.; et al. Vibriosis in aquaculture. In Proceedings of the 16th EAFP Conference, Tampere, Finland, 4 September 2013; pp. 138–148.
5. Veić, T. Characterization of Vibrio Species Isolated from European Sea Bass (*Dicentrarhus Labrax*, Linnaeus, 1758) Farmed on the Eastern Adriatic Sea. Master's Thesis, University of Zagreb, Zagreb, Croatia, 2016.
6. Rosado, D.; Xavier, R.; Severino, R.; Tavares, F.; Cable, J.; Perez-Losada, M. Effects of disease, antibiotic treatment and recovery trajectory on the microbiome of farmed seabass (*Dicentrarchus labrax*). *Sci. Rep.* **2019**, *9*, 18946. [CrossRef] [PubMed]
7. Rosado, D.; Perez-Losada, M.; Severino, R.; Cable, J.; Xavier, R. Characterization of the skin and gill microbiomes of the farmed seabass (*Dicentrarchus labrax*) and seabream (*Sparus aurata*). *Aquaculture* **2018**, *500*, 57–64. [CrossRef]
8. Bruijn, I.D.; Liu, Y.; Wiegertjes, G.F.; Raaijmakers, J.M. Exploring fish microbial communities to mitigate emerging diseases in aquaculture. *FEMS Microbiol. Ecol.* **2018**, *94*, 1–12. [CrossRef]
9. Reverter, M.; Tapissier-Bontemps, N.; Lecchini, D.; Banaigs, B.; Sasal, P. Biological and Ecological Roles of External Fish Mucus: A Review. *Fishes* **2018**, *3*, 41. [CrossRef]

10. Gatesoupe, F.-J.; Huelvan, C.; Le Bayon, N.; Le Delliou, H.; Madec, L.; Mouchel, O.; Quazuguel, P.; Mazurais, D.; Zambonino-Infante, J.-L. The highly variable microbiota associated to intestinal mucosa correlates with growth and hypoxia resistance of sea bass, *Dicentrarchus labrax*, submitted to different nutritional histories. *BMC Microbiol.* **2016**, *16*, 266. [CrossRef]

11. Perez-Pascual, D.; Estelle, J.; Dutto, G.; Rodde, C.; Bernardet, J.-F.; Marchand, Y.; Duchaud, E.; Przybyla, C.; Ghigo, J.-M. Growth Performance and adaptability of European sea bass (*Dicentrarchus labrax*) gut microbiota to alternative diets free of fish products. *Microorganisms* **2020**, *8*, 1346. [CrossRef]

12. Silvi, S.; Nardi, M.; Sulpizio, R.; Orpianesi, C.; Caggiano, M.; Carnevali, O.; Cresci, A. Effect of the addition of *Lactobacillus delbrueckii* subsp. delbrueckii on the gut microbiota composition and contribution to the well-being of European sea bass (*Dicentrarchus labrax*, L.). *Microb. Ecol. Health Dis.* **2008**, *20*, 53–59. [CrossRef]

13. Roquigny, R.; Mougin, J.; Le Bris, C.; Bonnin-Jusserand, M.; Doyen, P.; Grard, T. Characterization of the marine aquaculture microbiome: A seasonal survey in a seabass farm. *Aquaculture* **2021**, *531*, 735987. [CrossRef]

14. Kapetanovic, D.; Gavrilovic, A.; Jug-Dujakovic, J.; Smrzlic, I.V.; Kazazic, S.; Bojanic-Rasovic, M.; Kolda, A.; Pesic, A.; Peric, L.; Zunic, J.; et al. Assessment of microbial sea water quality and health status of farmed European seabass (*Dicentrarchus labrax*) in Eastern Adriatic sea (Montenegro and Croatia). *Stud. Mar.* **2019**, *32*, 52–64. [CrossRef]

15. Rosado, D.; Pérez-Losada, M.; Pereira, A.; Severino, R.; Xavier, R. Effects of aging on the skin and gill microbiota of farmed seabass and seabream. *Anim. Microbiome* **2021**, *3*, 10. [CrossRef]

16. Pepi, M.; Focardi, S. Antibiotic-Resistant Bacteria in Aquaculture and Climate Change: A Challenge for Health in the Mediterranean Area. *Int. J. Environ. Res. Public Health* **2021**, *18*, 5723. [CrossRef] [PubMed]

17. Damir, K.; Irena, V.S.; Damir, V.; Emin, T. Occurrence, characterization and antimicrobial susceptibility of *Vibrio alginolyticus* in the Eastern Adriatic sea. *Mar. Pollut. Bull.* **2013**, *75*, 46–52. [CrossRef] [PubMed]

18. Wilson, K.H.; Blitchington, R.B.; Greene, R.C. Amplification of bacterial 16S ribosomal DNA with polymerase chain reaction. *J. Clin. Microbiol.* **1990**, *28*, 1942–1946. [CrossRef] [PubMed]

19. Hall, T.A. BioEdit: A user-friendly biological sequence alignment editor and analysis program for Windows 95/98/NT. *Nucl. Acids. Symp. Ser.* **1999**, *41*, 95–98.

20. Tamura, K.; Stecher, G.; Kumar, S. MEGA11: Molecular Evolutionary Genetics Analysis Version 11. *Mol. Biol. Evol.* **2021**, *38*, 3022–3027. [CrossRef] [PubMed]

21. Philipose, K.K.; Sharma, S.R.; Loka, J.; Damodaran, D.; Rao, G.S.; Vaidya, N.G.; Sonali, S.M.; Sadhu, N.; Dube, P. Observations on variations in physico-chemical water parameters of marine fish cage farm off Karwar. *Indian J. Fish.* **2012**, *59*, 83–88.

22. Kapetanovic, D.; Smrzlic, I.V.; Valic, D.; Teskeredzic, Z.; Teskeredzic, E. Culturable microbiota associated with farmed Atlantic bluefin tuna (*Thunnus thynnus*). *Aquat. Living Resour.* **2017**, *30*, 30. [CrossRef]

23. Matijevic, S.; Kuspilic, G.; Morovic, M.; Grbec, B.; Bogner, D.; Skejic, S.; Veza, J. Physical and chemical properties of the water column and sediments at sea bass/sea bream farm in the middle Adriatic (Maslinova bay). *Acta Adriat.* **2009**, *50*, 59–76.

24. Roux, F.L.; Wegner, K.M.; Baker-Austin, C.; Vezzulli, L.; Osorio, C.R.; Amaro, C.; Ritchie, J.M.; Defoirdt, T.; Destoumieux-Garzón, D.; Blokesch, M.; et al. The emergence of *Vibrio* pathogens in Europe: Ecology, evolution, and pathogenesis (Paris, 11-12th March 2015). *Front. Microbiol.* **2015**, *6*, 830. [CrossRef] [PubMed]

25. Vatsos, I.N. Standardizing the microbiota of fish used in research. *Lab. Anim.* **2017**, *51*, 353–364. [CrossRef] [PubMed]

26. Franco, A.; Rückert, C.; Blom, J.; Busche, T.; Reichert, J.; Schubert, P.; Goesmann, A.; Kalinowski, J.; Wilke, T.; Kämpfer, P.; et al. High diversity of *Vibrio* spp. associated with different ecological niches in a marine aquaria system and description of *Vibrio aquimaris* sp. nov. *Syst. Appl. Microbiol.* **2020**, *43*, 126123. [CrossRef] [PubMed]

27. Onianwah, I.F.; Stanley, H.O.; Oyakhire, M. Microorganisms in Aquaculture Development. *Glob. Adv. Res. J. Microbiol.* **2018**, *7*, 127–131.

28. Čož-Rakovac, R.; Strunjak-Perović, I.; Topić Popović, N.; Hacmenjak, M.; Šimpraga, B.; Teskeredžić, E. Health status of wild and cultured sea bass in the northern Adriatic Sea. *Vet. Med.* **2002**, *47*, 222–226. [CrossRef]

29. Romero, J.; Feijoo, C.G.; Navarrete, P. Antibiotics in Aquaculture—Use, Abuse and Alternatives. In *Health and Environment in Aquaculture*; Carvalho, E., Ed.; InTech: London, UK, 2012; ISBN 978-953-51-0497-1. Available online: http://www.intechopen.com/books/health-and-environment-in-aquaculture/antibioticsin-aquaculture-use-abuse-and-alternatives-23/12/2021 (accessed on 23 December 2021).

30. Pujalte, M.J.; Sitjà-Bobadilla, A.; Macián, M.C.; Álvarez-Pellitero, P.; Garay, E. Occurrence and virulence of *Pseudoalteromonas* spp. in cultured gilthead sea bream (*Sparus aurata* L.) and European sea bass (*Dicentrarchus labrax* L.). Molecular and phenotypic characterisation of *P. undina* strain U58. *Aquaculture* **2007**, *271*, 47–53. [CrossRef]

31. Hamza, F.; Kumar, A.R.; Zinjarde, S. Efficacy of cell free supernatant from *Bacillus licheniformis* in protecting *Artemia salina* against *Vibrio alginolyticus* and *Pseudomonas gessardii*. *Microb. Pathog.* **2018**, *116*, 335–344. [CrossRef]

32. Vendramin, N.; Zrncic, S.; Padros, F.; Oraic, D.; Le Breton, A.; Zarza, C.; Olesen, N.J. Fish health in Mediterranean Aquaculture, past mistakes and future challenges. *Bull. Eur. Assoc. Fish Pathol.* **2016**, *36*, 38–45.

33. Lasa, A.; Avendaño-Herrera, R.; Estrada, J.M.; Romalde, J.L. Isolation and identification of *Vibrio toranzoniae* associated with diseased red conger eel (*Genypterus chilensis*) farmed in Chile. *Vet. Microbiol.* **2015**, *179*, 327–331. [CrossRef]

34. Zorrilla, I.; Chabrillón, M.; Arijo, S.; Díaz-Rosales, P.; Martínez-Manzanares, E.; Balebona, M.C.; Morinigo, M.A. Bacteria recovered from diseased cultured gilthead sea bream (*Sparus aurata* L.) in southwestern Spain. *Aquaculture* **2003**, *218*, 11–20. [CrossRef]

35. Kang, C.H.; Kim, Y.G.; Oh, S.J.; Mok, J.S.; Cho, M.H.; So, J.S. Antibiotic resistance of *Vibrio harveyi* isolated from seawater in Korea. *Mar. Pollut. Bull.* **2014**, *86*, 261–265. [CrossRef] [PubMed]

36. Parin, U.; Erbaş, G.; Savasan, S.; Yuksel, H.T.; Gurpinar, S.; Kirkan, S. Antimicrobial resistance of *Vibrio* (*Listonella*) *anguillarum* isolated from rainbow trouts (*Oncorhynchus mykiss*). *Indian J. Anim. Res.* **2019**, *53*, 1522–1525. [CrossRef]

37. Kummerer, K. Resistance in the environment. *J. Antimicrob. Chemother.* **2004**, *54*, 311–320. [CrossRef] [PubMed]

38. Chelossi, E.; Vezzulli, L.; Milano, A.; Branzoni, M.; Fabiano, M.; Riccardi, G.; Banat, I.M. Antibiotic resistance of benthic bacteria in fish-farm and control sediments of the Western Mediterranean. *Aquaculture* **2003**, *219*, 83–97. [CrossRef]

39. Baquero, F.; Martinez, J.-L.; Canton, R. Antibiotics and antibiotic resistance in water environments. *Curr. Opin. Biotechnol.* **2008**, *19*, 260–265. [CrossRef] [PubMed]

40. Kostelic, A.; Artukovic, B.; Beck, R.; Benic, M.; Cergolj, M.; Stokovic, I.; Barac, Z. Diseases of sheep on Croatian islands. In Proceedings of the XVI. Congress of the Mediterranean Federation for Health and Production of Ruminants: FeMeSPRum, Zadar, Croatia, 26 April 2008; pp. 227–232.

41. Official Gazette 21/2011 Ordinance on Pharmacologically Active Substances and Their Classification in Relation to the Maximum Permitted Levels of Residues in Food of Animal Origin. Available online: https://narodne-novine.nn.hr/clanci/sluzbeni/2005_03_29_510.html (accessed on 23 December 2021).

42. Done, H.Y.; Venkatesan, A.K.; Halden, R.U. Does the Recent Growth of Aquaculture Create Antibiotic Resistance Threats Different from those Associated with Land Animal Production in Agriculture? *AAPS J.* **2015**, *17*, 513–524, Erratum in: *AAPS J.* **2016**, *18*, 1583. [CrossRef] [PubMed]

MDPI
St. Alban-Anlage 66
4052 Basel
Switzerland
Tel. +41 61 683 77 34
Fax +41 61 302 89 18
www.mdpi.com

Journal of Marine Science and Engineering Editorial Office
E-mail: jmse@mdpi.com
www.mdpi.com/journal/jmse